MATTER AND METHOD

MATTER
&
METHOD

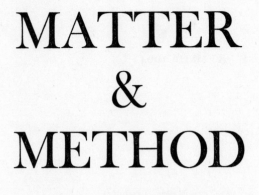

R. Harré

Lecturer in Philosophy of Science
University of Oxford

London
MACMILLAN & CO LTD
NEW YORK ST MARTIN'S PRESS

MACMILLAN AND COMPANY LIMITED
St Martin's Street London WC 2
also Bombay Calcutta Madras Melbourne Sydney

THE MACMILLAN COMPANY OF CANADA LIMITED
70 Bond Street Toronto 2

ST MARTIN'S PRESS INC
175 Fifth Avenue New York 10 NY

PRINTED IN GREAT BRITAIN
BY R & R CLARK LTD EDINBURGH

CONTENTS

PREFACE

Philosophers learn the doctrines of Locke and Berkeley about matter and qualities. Very rarely do they come to see them in the setting in which they were composed, the development of a new way of looking at the world. In this book an attempt is made to show a thread of argument which runs through both the scientists and philosophers of the Seventeenth and Eighteenth Centuries, each contributing to the establishment of the Corpuscularian Philosophy. I hope that by following this thread students of philosophy will get a more just view of the British Empiricists, and historians of science a better understanding of the involvement of conceptual change in scientific discovery.

INTRODUCTION

There are so many different activities to which people give the name Philosophy that it is quite essential nowadays to say which one of these one is pursuing and so stifle some criticism by anticipating it. Two philosophical pursuits will be found exemplified in this study of matter: each supporting the other. One is critical philosophy and the other constructive philosophy. Critical philosophy originates when, by philosophical argument, something to which we are all prepared to assent is apparently shown to be impossible. By careful analysis both of the unreflective opinion and of the philosophical argument one hopes to resolve the paradoxes that philosophising generates. A famous example of this kind of situation is the 'problem' of Mind and Matter. Everyone knows that he can physically carry out decisions which he has made by going through some mental process. By a philosophical argument one can be led to ascribe physical happenings to a material substance and mental happenings to a spiritual substance which have nothing whatever in common. How then can they interact? We easily pass to the paradoxical conclusion that the everyday activity of carrying out decisions is impossible.

In resolving difficulties such as the one I have just mentioned, we come upon certain fundamental concepts like *matter* and *mind*, *cause* and *action* and the like and we are driven by the necessities of the resolution of philosophical paradoxes to give some account of these concepts. This leads to attempts at constructive philosophy. Constructive philosophy is like an experiment: having made some suggestion about the way we should understand a fundamental concept we then try to fit the clarified version into all the relevant linguistic environments of the concept unrefined, and we test it against the accepted facts which that concept is used to express. (Sometimes of course what counts as a fact is a function of the way we understand the concept.) Contradiction and disparity count as refuting the thesis that the concept is as we have supposed it to be for the purposes of the critical analysis. In this study the concept upon which our 'experiment' is to be conducted is that for which the words *substance* and *matter* are commonly used. We shall find that not only does our way of understanding the concept determine our views of the 'ultimate nature of reality' but also influences the way we approach the world in seeking knowledge of its constitution and behaviour. The way we understand the concept of matter is closely connected with what we find acceptable as legitimate scientific method.

Part 1

THE ANALYSIS
OF CONCEPTUAL
SYSTEMS

The Scope of this Study

The aim of science is to provide us with satisfactory explanations of the actual course of nature. Prediction and control follow. But to make this aim clear is more difficult than stating it, since a number of difficult questions need to be answered before this aim can be understood. What are explanations and how do they work? What criteria do explanations need to satisfy before we are ready to accept them as satisfactory? When does explaining come to an end? These are some of the questions which are answered by the study of the logic of scientific discourse. What are the happenings which science explains? What is the world in which these occur? How is this world related to us? These are some of the questions of what one might call the metaphysics of science. One might be tempted to think that the questions in this latter group are easily answered, and metaphysics disposed of, by directing the questioner's attention to the rising of the sun, to freezings and meltings, fevers and the sound of bells; to people, stones and sunsets, and saying: 'These are the happenings and this is the world and science is the explanation of it'.

In a sense one can hardly deny that this sort of answer is correct. Scientists, however, do not deal exclusively in what we can perceive. They also deal in planetary systems, atoms and molecules and energy, viruses and sound-waves, photons and waves in the 'ether'. Our instruments and their behaviour are among the things we can perceive, but quite frequently what we say they are detecting cannot be perceived. We are quite ready to say that there is an insensible cause of the visible effect; and then open a new branch of science by studying it. What our instruments read and what we say they are detecting

3

are quite clearly different. Is there the same difference between what we perceive and what causes our perception? By assimilating these cases great philosophical difficulties have been created for those who wish to maintain that scientists do actually discover real processes in nature which manifest themselves to us only through their effects upon us and upon our instruments. 'If an electron has passed through a geiger counter then the counter clicks' is the kind of statement that is used to link instruments and perceptions to the postulated entities of the world. Is it just a mistake in logic to argue that from the occurrence of clicks we can infer the existence of electrons causing the clicks? This is one sort of difficulty for the metaphysics of science. There is another.

We know that our perceptions are, in part, the result of the mental ordering and grouping of our sensations. It is an easy step from this fact to sceptical doubts about the degree of resemblance which we are entitled to assume that our perceptions bear to those things, properties and processes which we want to say cause our perceptions. There are all sorts of responses to such doubts. We might insist on the reality of our sensations, or our perceptions, or the entities of science, or any combination of these.

Phenomenalists, searching for indubitable truths, have been inclined to find them in assertions about the copresence of sensations in the sensory fields of individual persons. This attitude makes the reality of what we usually say we perceive and its unperceived causes doubtful in more or less degree. There is then a problem for phenomenalists to solve if they wish to pass from the evidence of their senses to what we naively suppose to be evidently their causes.*

Idealists delete the problem by pushing the phenomenalist position to its ultimate conclusion. If for an idealist 'to be is to be perceived' then perceptions are ontologically dependent on the perceiver and there is no independent, causally active

* For a very full discussion of this position as it affects the philosophy of science, see P. Alexander, *Sensationalism in Science*. Routledge & Kegan Paul, 1963.

world with which we interact. But it seems that, without God to teach us the laws of nature, the universe on this view fragments into closed individual worlds without the possibility of communication. In Berkeley and Leibniz one can find the position developed. But being as implausible as it is irrefutable we shall not take it as either a starting-point or a goal to be achieved.

Platonists, taking the distinction between perceptions and their causes more seriously, and finding knowledge of the causes at the ends of aesthetically uplifting chains of reasoning, were inclined to find reality in the hidden generators of the natural world we perceive, while we, the perceivers, are like so many cave dwellers who, with their backs to the light, see only shadows of the beauties outside.

An empiricist might be characterized as one who believes in it all. For him the sources of knowledge are sensory. There are sensations and they are bound together by the cement of experience into percepts. *The critical study of percepts is what enables us to pass to the causes of what we perceive;* and though knowledge of this kind may be good it can never be certified beyond the possibility of revision.

Of all philosophical positions empiricism accords most closely with the assumptions under which, with the possible exception of the work of some modern theoreticians in physics, the natural sciences have been pursued. One can go farther than this I believe by developing arguments that suggest that the development of natural knowledge would have been seriously impaired had any other philosophical theory been incorporated in the assumptions of the seekers after natural knowledge. Indeed, in those cases where other philosophical theories have affected science the result has been stultificatory or if progress has been made, it has been despite philosophical difficulties. It is then to varieties of empiricism that we shall direct our attention, taking the arguments for the possibilities of empiricism for granted.

In this present study my aim is to exhibit the role of one

specification of the empiricist view in the structure of science. This is the doctrine of *matter*. All that exists is matter, and matter is what exists. One way out of this apparently circular run of definitions is to try to define matter in some special way, and it is to the study of one of these ways, matter defined by mechanical behaviour, that I shall devote most attention. I hope to demonstrate in the course of this study that the methods of investigation held to be proper and the kinds of explanations held to be appropriate in any given epoch depend, at least in part, on the particular specification of the general concept of matter popular in that epoch.

This raises another problem. At any given epoch there is an end to explanation, and this end is reached when the entities and processes of an explanatory mechanism are those which form part of the denotation of what might be called the *general conceptual system* of the epoch. The plan of this book is determined by two questions which this suggestion raises.

(i) What is a general conceptual system, and how does it influence the construction of science?

(ii) What sort of arguments can be devised to recommend a particular choice of general conceptual system, namely the choice of a particular specification of the concept of matter as the fundamental concept; and of the so-called primary properties as defining it.

Both parts of the plan devolve round the science and philosophy of the seventeenth and eighteenth centuries. This should not be taken to suggest that this is a purely historical study. History provides the raw material for analysis, but what can be learned from an investigation of this kind is, one hopes, of general significance for the understanding of science, at whatever epoch one considers it. I shall, therefore, feel no compunction whatever in tidying up people's thoughts and in drawing out hints into philosophical arguments; in the confidence that such a-historical activities are not unhistorical in a wider sense.

Matter and Metaphysics

In some important way it is being material that distinguishes the real from the fictitious and the imaginary. Things, we are strongly inclined to say, are made up of matter. We are ready to speak of the conservation of matter, of the equivalence of matter and energy. Physics traditionally begins with the properties of matter. Matter, we are quite ready to say, we know to be elastic, to have inertia, to be attracted by other matter, to resist penetration and so on. But matter is like the demon who treads close behind us in the dark: there when we do not look at it but straightway disappearing when we do. When a materialist says that only matter is real what, if anything, is he saying? In trying to find an answer to this question I want to search out those concepts of matter that are embedded in the classical physical sciences. At the same time it will be necessary to give some account of what a fundamental concept is, what the general conceptual system of an epoch is and how it functions in providing the basic materials for scientific thought, for, if anything, matter is such a concept and figures in such a system. The investigation of the concept of matter can also serve as a step in the unravelling of the structure of the intellectual part of science, because matter played such an important part in two great conceptual systems of the past, the Aristotelian and the Newtonian. Perhaps in this way we may be freed from the tendency there still seems to be, to carry through into modern physics prejudices as to what there *is* which are really disguised refusals to give up a doctrine of matter that developed only in the seventeenth century.

There are two main concepts of matter:

(1) *The essentialist doctrine of matter*: that some set of the properties of things are the essential propreties of matter and thus

B

constitutive of it. Any continuously existing region of space–time exhibiting these essential (or primary) properties is a material thing. This is the doctrine whose refined form was the product of the empiricism of the seventeenth and eighteenth centuries.

(2) *The invariants doctrine:* that processes do not have continuity just by accident, but their continuity is a manifestation of an underlying unity which persists through radical changes of many of the properties of things. The invariants in natural processes are the items which are material. The Aristotelian doctrine of prime matter or substance, that of which all qualities are properties but which has none itself, is the *reductio ad absurdum* of the invariants doctrine of matter.

The problem with which we shall be concerned is the study of the way one particular essentialist doctrine of matter, the Newtonian, came to be held; and in doing this we shall come to gain some idea of large-scale conceptual change in science, of which this doctrine and its development is an outstanding example.

Theories

The prime piece of intellectual equipment is the theory. We shall find it necessary to recognize two radically different kinds of theory, each playing an indispensable role in the sciences. Logicians, having tidy minds, have been tempted to try to reduce the two kinds to one. By presenting the two kinds of theory in some detail, their differences, in both form and function, should emerge clearly. Roughly speaking, there are theories which *connect up* sets of facts all of the same kind (what it is for two facts to be of the same kind will emerge from the discussion); and there are theories which *explain* one set of facts by adducing in explanation another and different set of facts. Those theories which connect up but do not explain, except in a minimal sense, I shall call *reticular* theories, and those which explain in a maximal sense, *explanatory* theories.

RETICULAR THEORIES

A characteristic reticular theory is simple Newtonian dynamics. Newton knew of a number of regularities in the phenomena of motion, some from the investigations of Kepler and Galileo, and some from discoveries made in his own time. His problem was essentially to find a way of connecting up *all* the facts of motion, both terrestrial and celestial, partially and fragmentarily connected up by his predecessors.

Newton's method, looked at from a logical point of view, was to introduce a concept *force*, which would, with certain rules of combination, link up into a connected whole the isolated facts of motion. For example, how to find a way of connecting up the motions of two bodies before and after impact? Newton's way was to define *force* as mass-acceleration, then provide the required connectedness by the Principle of Action and Reaction; this is that in impact the force which each body exerts upon the other is identical in magnitude but opposite in direction. Symbolically we have:

$$F = -F' \qquad \text{(Action and Reaction)}$$
$$ma = m'a' \qquad \text{(Definition of force)}$$
$$\frac{m(v-u)}{t} = \frac{-m'(v'-u')}{t} \quad \text{(Definition of acceleration)}$$
$$mv - mu = -m'v' + m'u'$$
$$mv + m'v' = mu + m'u'$$

These transformations lead us to a relation expressed in terms of a new concept *momentum*, in which total momentum after impact is equal to total momentum before impact, and momentum is given entirely in terms of properties which are measurable (though mass is measurable only indirectly).

To be quite clear about the role of force in this system we need to look more closely at what is to count as a fact of dynamics. We can easily calculate a numerical value for the force which we say is acting when a given body undergoes some given acceleration. But it is equally clear that to say that there is the force and that it has this value is not to assert the obtaining of some fact, in the same sense as the assertion that the body traverses a certain distance in a certain time is to state a fact. What we can observe and measure are distances, times and motions. We can measure forces but not observe them, and to do this we calculate the value of the force from the space-time changes that a force produces in a given body. If we say that the facts of dynamics are given by descriptions of processes in terms of concepts derived directly from observation, that there are forces acting and that they have such and such magnitudes are not facts of dynamics. To make this distinction sharp we have to specify more exactly what we are to understand by the 'direct derivation' of a concept from observation. The values of the magnitudes of forces are certainly derived from observation, but we want to say that by contrast with the magnitude of distances and times they are not directly derived.

One difference between direct and indirect derivation lies in this: that a scientific concept is directly derived from observation if its immediate referent is capable of being perceived, e.g. as the referents of all colour words are capable of being perceived. A concept is indirectly derived when its immediate referent is not capable of being perceived, but is the product of an intellectual operation, such as the definition of a concept as some function of concepts directly derived. We must be careful not to misunderstand this distinction. Though *we* can feel forces when we push and bend things, and in consequence see them move, even though we see a similar movement when an engine pushes a truck or a weight bends a bar we feel nothing. I have said that the facts of dynamics are expressed in concepts which are used in the description of properties or processes one can perceive in nature, but that in their use in dynamics such concepts are employed in a refined form. The refinement of observational concepts for scientific purposes takes place by a process of *reduction*.

Consider the concept of *speed* and its refined form *velocity*. The concept of velocity is reached from the concept of speed by an intellectual process which is, in part, determined by the requirements which have to be met in order to give a numerical value to the magnitude of spatial translation and rotation. To express the rate of change of place with respect to time with any degree of accuracy we measure the distance covered and the time taken to cover that distance.

$$\text{Speed} = v \text{ feet per second,}$$

expresses a measure of speed in terms of the measurements required to give a numerical value to the magnitude of the change of place with time. It is now a short step from thus defining a measure of speed to redefining the concept along the same lines. If we define a refined concept of change of place with time as

$$\text{Velocity} = \frac{\text{Distance}}{\text{Time}}$$

and
$$\text{Instantaneous Velocity} = \underset{t \to 0}{Lt} \cdot \frac{\delta s}{\delta t}$$

we have replaced the raw concept *speed* with the precise concept *velocity*, which is necessarily measured in units of distance and time, such as feet per second.

But when we say mass-acceleration is a measure of force this is not a refinement of an observational concept, nor a reduction of that concept to some function of independent concepts, it *is* the concept of force as it functions in dynamics. To say that force is mass-acceleration is to propose a construction for a new and wholly theoretical concept. It might be objected that though we cannot observe a naked force, we have, in the

spring balance, a device for measuring force without the intermediate mass-acceleration. But this is to misunderstand the
nature of the spring balance. It is to assimilate it to instruments of the type of the foot-rule, whereas it is actually an
instrument of the type of the kilowatt-hour meter. A spring
balance is essentially a computer, albeit of a very simple type,
and depends upon Hooke's Law for its effectiveness as an
evaluator of forces. Another way of putting this point is to
observe that the units of force, poundal and pound-weight,
are not determined by instruments for measuring forces, but
are derived from concept-complexes used for calculating
forces. Here we contrast the foot-rule, which is not only an
instrument for measuring lengths, but is itself a lengthy body
which could, if we chose, play the role of a standard.

   It would be too simple to say that a refined form of a concept is determined just by the necessities of measurement. It is
in part determined by an ideal of conceptual unity. It has
been an ideal of scientists to try to carry out their refinements
and reconstructions of concepts in terms of the same set of
fundamental concepts. Galileo's mysterious difficulties with
the concept of acceleration can, it seems to me, be put down
to a tension between the above ideal and a certain standard of
theoretical elegance set by the geometrical mathematics to
which he was confined. Our ordinary concept of acceleration
is hazy and undefined. We vacillate between an increase of
speed with distance and an increase of speed with time.
Motorists tend to adopt the former sense in discussing the
performance of their machines. It must be emphasized that
both these are legitimate concepts, but of course they differ
in the consequences which follow their adoption. Suppose
we call time-acceleration 'a' and distance-acceleration 'd';
and represent change in velocity as 'Δv', change in distance
as 'Δs', change in time as 'Δt'. Then

$$a = \frac{\Delta v}{\Delta t}$$

and this definition requires equal increments of speed for equal time increments. But if we choose

$$d = \frac{\Delta v}{\Delta s}$$

(as Galileo did originally), we require equal increments of speed for equal space increments. Now comparing uniform velocity with uniform acceleration we get a kind of disparity which Galileo's mathematics was not equipped to handle. One kind of change, velocity, was treated as change with respect to time, and the other kind of kinematic change, acceleration, was treated as change with respect to distance. Why did he choose the motorists' concept of acceleration? I think the answer must lie in the remnants of the medieval doctrine that a spatial interval can represent only a spatial interval; that geometry is an ideal picture of the world and bears a pictorial relation to it. In 1638, by plumping for the ideal of conceptual unity, in this case that all changes are understood with respect to time, Galileo's difficulties were over. What reduction he was to make was not given to him, it was not to be discovered just by an inspection of his conceptual materials, for he was driven this way and that as one or other theoretical ideal was uppermost in his mind.

A reticular theory can be defined as a set of relationships between refined observational concepts, mediated by one or more theoretical concepts which are to be understood wholly in terms of a complex of the refined observational concepts of the theory. Kinematics, dynamics, matrix mechanics, geometrical optics etc. are all reticular theories. Reticular theories, for all their importance and practical value, do not explain phenomena. To show that this is so we need only work out what sort of causation they imply or presuppose between the phenomena they serve to order.

Philosophers have recognized both a minimal and a maximal sense of causation. The maximal sense is pretty much the sense we give to causation in unphilosophical uses of that concept, while the minimal sense is very much a philosophers' concept, the result of a critical, logical analysis, notably by Hume, of the maximal sense of causation.

The maximal sense can be seen in some examples:

Q. What causes dough to rise?

A. Yeast *produces* carbon dioxide and this *produces* a sponge-like consistency in the mixture. . . .

Q. What causes the hands of a clock to move?

A. The spring tension which is *transmitted through* the movement. . . .

Three features of these accounts should be noticed.

(i) In the maximal sense of causation there is the idea of agency or productivity of the effect by the cause — the cause together with the conditions in which it acts generates the effect.

(ii) Each answer is open, since we can continue to expand it by giving more details of the mechanism by which the causal agent produces the effect. For instance we can fill out the dough story with talk of carbohydrates and enzymes, solution pressures and so on; and we can continue to specify more and more details of the clock movement, the escapement, the gear trains and so on.

(iii) Each answer is a list of necessary conditions for the occurrence of the effect, or at least each answer can be treated in this way, for each element in an answer mentions a feature of the situation which is of such a kind that if it were absent the effect would not occur. But no answer, however expanded, contains sufficient conditions for an effect, since the list of necessary conditions is open. For instance the temperature, humidity and atmospheric constitution must be right before the activity of the yeast will cause the dough to rise; and consideration of the conditions for those conditions to be right opens a regress of conditions which cannot be closed short of the whole past history of the universe. The set of conditions which must obtain for any given spring to drive any given clock is open in a similar way.

To sum up this maximal sense of causation we suppose:

(a) that the cause has some connection with the effect, by which it generates the effect;

(b) that some selection of the set of necessary conditions (which would be sufficient only if extended indefinitely) can be made which will suffice for the practical production of the effect against a background of relative stability in all other conditions.

The minimal sense of causation is arrived at by the following argument:

(i) The cause-event or phenomenon is an observable occurrence and so is the effect-event or phenomenon; but the relation between them is not an observable event or occurrence. This follows from a general philosophical position expressed in the principle: 'All we can observe in a serial change are the events which severally compose it'.

(ii) The connection relation must therefore be a wholly conceptual one, and is, according to Hume, nothing more than the expression of a habit of mind ingrained by the invariable concomitance of events of the cause type with events of the effect type.

It has often been claimed that even if the minimal sense of causation is not that used in the natural sciences, it ought to be. We can now see that reticular theories fit in with the minimal concept of causation perfectly. The connection relation between phenomena can only be concomitance, since the relation concept, e.g. *force*, is, by the nature of its logic, wholly theoretical. Momenta are related by equal and opposite forces, but this is a wholly conceptual and not an observable relation, since forces are theoretical concepts, never observed in nature. However, it might be objected on behalf of a maximal causal interpretation of reticular theories that we are strongly tempted to say, for instance, that force *produces* acceleration. To put this point in another way: does not the effectiveness of the force concept in reticulating the observable phenomena of dynamics justify our concluding that there really are forces which generate motions? To answer the objection it could simply be insisted that the reification of forces is wholly unjustified, for all we observe in impacts and gravitations are motions. The action and the reaction, the gravitational forces, are not observed. But this would hardly meet the case, since the objection is the stronger one that the logical power of the theoretical concept is such that we are forced to acknowledge the existence of a referent for it whether or not this referent can be observed. It might be argued that a powerful criterion for the existence of a referent for a concept is a successful employment of that concept in reticulation of observed phenomena. That this compulsion is essentially anthropomorphic in origin can be demonstrated.

The concept *force* can be eliminated from dynamics, without any loss of empirical content. One method of elimination is that due to Mach.*

Roughly, Mach's elimination amounts to this: every mechanical phenomenon is an interaction, and every mechanical problem is reducible either to the task of finding the velocities, accelerations and positions of masses after an interaction, given the values of these parameters before, or to the complementary problem of working back to the initial state of some system of interacting bodies. In the Newtonian mechanics the laws linking before and after situations are constructed by asserting that in every mechanical interaction the Action and Reaction principle holds. This principle states that in every mechanical interaction there are a pair of equal and opposite forces. Rules for calculating these forces can be derived from Newton's Second Law, and the occasions of their use, that is, occurrences which are to count as mechanical interactions are specified by Newton's First Law, that is, mechanical interactions occur when a body changes its state of rest or uniform motion. But since Action and Reaction are equal and opposite,

* Mach, E., *The Science of Mechanics*, Ch. 2, § v.

force never appears in the answer to a genuine mechanical problem. It is only part of the working. Concepts which are only part of the working are not used to express facts.

However, force is not similarly eliminable from, say, instructions on how to ride a bicycle, without loss of empirical content. In such instructions we use the concept of force explicitly to refer to certain sensations experienced by the rider. But it is wholly anthropomorphic, given the formal eliminability of force from dynamics, to claim that the concept has existential reference in those systems in which no organic material is immediately involved. We cannot rationally make reference to the 'sensation experienced by impacting bodies or gyrating planets', unless it should be someone's misfortune to be one of the bodies involved in the impact, or to be a lost astronaut. Compare, for instance, knowing that a man, a mouse, a loaf of bread, a tin can, will, relative to some satellite in which these items are orbiting, be weightless, with what being weightless feels like to the man and to the mouse.

One final point about reticular theories. They are what I shall call 'closed conceptual systems', that is additional theoretical concepts must be definable in terms of the original concepts of the theory. For example, if $F = df \cdot ma$, and every relation between dynamic observables is mediated by equality of forces, as in the action and reaction principle, then every additional dynamic concept is some function of F, i.e. of ma. For instance, Work (or Energy) is Distance $\times F$, Impulse is Time $\times F$, and so on. Hence

$$\text{Work/Distance} = \text{Impulse/Time}$$

i.e. $$\text{Work} = \text{Distance} \times \text{Impulse/Time}$$

and so on. The system of laws can, in this way, be made to look like a string of tautologies. Similar, though more complex, mutual definitions could be constructed for geometrical optics between such concepts as refractive index, focal length etc. This feature of reticular theories is not something to be deplored, or to be puzzled about. It is simply a consequence of the fact that the observables of the theory are related, and are *all* related, by a purely formal concept or concepts.

EXPLANATORY THEORIES

In contrast, explanatory theories are such that every relation among observables is mirrored by a corresponding relation among theoretical concepts. Theoretical concepts in explanatory theories are not used to form relations, as are the theoretical concepts of reticular theories, but to explain those relations which are already known. Explanatory theories are therefore logically more complex than reticular theories and, as we shall see, depend upon a richer notion of causality. They must contain at least as many relations among theoretical concepts as they explain between observables, together with sufficient relations between descriptive and theoretical concepts to enable a theoretical relation to be constructed for every member of the set of observable relations, expressed in terms of descriptive concepts.

For instance, calorific facts are expressed in various relations between the temperatures, masses, and so on of the objects of calorific change, the reticular theory of heat being formed by the basic 'Newtonian' law *Heat lost = Heat gained*, as widely known a law as the equality of Action and Reaction. These calorific facts are explained by theories in which the concepts temperature, mass etc. do not figure. They were explained by Carnot as the effects of the flow of the fluid caloric; by Clausius and Maxwell as the effects of the motion of molecules. The very same calorific facts were explained by two quite different theories. We could assimilate such theory construction to the model of the reticular theory only by the gross misrepresentation that 'caloric' and 'energy' are formally isomorphic in the structures of the theories. Similarly the chemical reactions:

> 'Heating copper in air gives copper calx'
> 'Heating copper calx with coal yields copper'

were given two quite different theoretical explanations, one in terms of phlogiston by Stahl, and the other in terms of oxygen by Lavoisier. The very same chemical facts were explained by two quite different theories.

In both Stahl's and Lavoisier's explanations there was a transfer of substance operative in the transformation of the reagents. For Stahl, heating copper in a metallic state drove off the fiery matter, phlogiston, leaving dephlogistated copper or calx. Coal leaves little ash when burnt so must lose a good deal of its substance in the form of phlogiston when it is heated, so before burning it must be a substance rich in fiery matter. This suggests that heating calx, deficient in phlogiston, with phlogiston-rich coal should yield copper. This hypothesis was confirmed by experiment. It must be emphasized that the whole string of experiments were qualitative, since the chemistry of Stahl's day was the experimental study of qualitative changes in matter. Lavoisier, having Priestley's experimental separation of dephlogistated air in mind, devised the contrary oxygen hypothesis to account for the phenomena. In calcination it is alleged, a gaseous matter is absorbed and in reaction with coal or any other reducing agent the same gaseous matter is given up by the calx. No qualitative difference would emerge whichever hypothesis was adopted, but from Black and Cavendish Lavoisier had the novel idea of a quantitative study of chemical reactions and, as we all now know, in quantitative experiments a difference between the hypotheses did show itself.

If we call the set of observable relations {A} and the set of theoretical relations {B}, then we must note the following points:

(i) {A} once discovered remain the same, while {B} can change radically.*

(ii) {A} concepts are fixed by observation or instrumentation while {B} concepts are more or less freely modifiable. (There are exceptions to this rule which we shall consider

---

* This does not imply that the members of {A} are eternal or quite incorrigible.

later.) The upshot of this is that {A} and {B} are, in a certain sense, independent.

In explanatory theories the links between theoretical and observable concepts are of two kinds, one causal and the other what I shall call, for want of a better term, *modal*.

Causal links. Field strength is linked to the acceleration of a charged particle in that field, as cause is to effect; the increase in average momentum of a gas causes the pressure to increase. The theoretical concept can be taken to refer to an event or process which causes the observable event or process as its effect.

Modal links. Temperature is linked to average kinetic energy of the molecules of a gas; wavelength of light is linked to the colour seen when the light impinges on the eye. Here it would not be correct to say that the increase in the average kinetic energy of the molecules causes the temperature to rise, or that light of longer wavelength causes us to see red. The reception of light of that wavelength and the seeing of the colour by the organism are two different aspects of the same phenomenon, just as the temperature and the average kinetic energy of molecules are also two different aspects of the same phenomenon.

Now it should be clear that in those cases where explanatory theories are linked to the phenomena to be explained by causal relations, this must be causality in the 'strong' or genetic sense. It cannot be causality in the 'weak' or concomitance sense, for we cannot observe a concomitance between the phenomena postulated by the theory and the phenomena observed, since, at least in the initial period of a theory's history, the former cannot be observed. This matter is of such importance for the analysis of the sciences that we must investigate it in some detail.

Production and Generation

The maximal sense of cause, in contrast with cause as regular concomitance, involves a sense of agency in the cause, by virtue of which it produces or generates the effect. Of causal agents we use the expressions 'produces' and 'generates', and these, as we shall see, are different concepts. A logical watershed between them can be demonstrated easily enough. We talk about the generation of electricity, and only rarely about the production of electricity; whereas we talk about the production of butter or coal or motor-cars, and never about the generation of butter, coal or cars. The obvious hypothesis which would account for this difference in use is that we use production for that which arises by the transformation of material substances, e.g. butter from dairy fat, cars from ingots and steel sheets; or by the separation of one substance from another, e.g. coal from its earthy matrix, gold from its ore. In both cases there is substance identity throughout the process of production.

We use 'generation' for those products which are different in kind from their antecedents. Here one should notice that 'product' is used to cover both the results of productions and generations, so the plain language is not an adequate guide to the logical distinctions required here. In the case of productions we have noticed that there is an identity of substance through the process. Similarly, in generations there is an identity which carries through from cause to effect but it is not an identity of substance. It is an identity of mechanism, of the physical system which, stimulated by that event we call the cause, generates the effect. For instance, 'Heating a body causes it to emit light', for the heating energizes the atoms of the substance and in relapsing to their ground state they emit light. Here the atoms are machines which, when stimulated by the cause, act in such a way as to generate the effect.

Throughout their history the sciences have tended to treat generations as cases of hidden production. The discovery of an apparent generation has led to a search for a substance or substance-like something which could be transformed in a causal sequence rather than allow the effect to be a generation. In the last example, for instance, we are strongly inclined to treat the process not as the generation, *de novo*, of light, but as the transformation of the same quantity of energy into a new form, that is, the production of light from heat is treated along the same lines as the process by which steel ingots are transformed into cars.

Incidentally what are we to make of the positivist, regularity 'theory' of causation in the light of the foregoing discussion? Hume proposed the regularity theory against the necessity theory, but it is not clear what sort of theory was being proposed, nor quite what sort of dispute has since been carried on about causation. Hume might have been advocating the analysis of the concept of cause as we actually use it, or he might have been recommending a new way in which we might use it. Clearly it is not an analysis of the concept of cause as it is actually used, for regularity of succession is no more than a criterion for asserting causal connection between events, and certainly is not what the assertion of causal connection means. If it is a recommendation as to how we should understand causal connection that we should identify the meaning of the expression with one particular criterion for asserting that the relation ' . . . causes . . . ' holds, a great deal of argument would be needed to persuade us to make such a drastic change. We would be required to understand the causal relation, not as asserting a connection between events or states in virtue of which one produces or generates the other, but as a purely external relation between events or states, as mysterious and wayward as if the sciences had never been invented. This would so drastically change the concept that we should be unable to use it for those relations of states or events which we know to be connected by some intervening mechanism or identity of substance, and we should simply have to invent a new word to mark the distinction between causal and non-causal regularities.

This brings us back to the matter of the last section. We can now distinguish two ways in which a system of states B can explain some other system of states A.

(*a*) Events and processes in B are modally linked to events of processes in A, that is, B events or states are identical with A events or states, only they are looked at from different points of view. For instance, what is a headache to me is an electrical disturbance to a physiologist, what is temperature to anyone considering the behaviour of gas samples is the average kinetic energy of molecules to someone considering the gas as a swarm of corpuscles and so on. It is an undue concentration on this kind of explanation that has given rise to the idea that an explanation is just a redescription of the phenomena to be explained.

(*b*) Events or processes in B either produce or generate events or processes in A, that is, B-states are different from A-states, but the former cause the latter. For instance, the rotation of the hands of a watch is a different process from the unwinding of the spring, the flow of electricity in a conductor is a different process from that conductor cutting through a magnetic field; but the unwinding of the spring causes the rotation of the hands and hence explains it, and the intersection of the field by the conductor causes the current. A particularly interesting case of this kind of explanation is the gene theory of inheritance. Between the characteristics of adult plants there are certain statistical relations through the generations which we call Mendel's Laws. The explanation of these relations is achieved by connecting the replication of nuclei of cells with the adult characters through the code theory of the nucleic acids. It has recently been discovered that there is a causal relation between triads of certain bases on the long chain molecules of the nuclear material and the characters of adult organisms. The base triads are not the adult characters considered from another point or view, in particular it would be wildly absurd to call them the adult characters redescribed!

C

The Hierarchy of Mechanisms

If we consider a whole science along the lines set out in the last section we find that it is possible to analyse it in such a way that it exhibits a hierarchical structure.

EXAMPLE I

The transformations of substances that are described in test-tube chemistry, what we call chemical reactions, can be described without reference to any specific theory. We can, for instance, say: 'muriatic acid and zinc give zinc salt and hydrogen gas' and all our terms muriatic acid, zinc, zinc salt and hydrogen gas can be made intelligible in terms of observable properties and reactions of these substances alone provided we ignore the etymological hinting of such words as gas. In just the same way we can successfully follow a recipe for making a cake without knowing anything about the substances we combine except how to recognize them on the shelves.

The atomic hypothesis and the theory of valency and chemical bonding enable us to account for just those reactions which do take place and to some extent to predict others which we can later observe. We are thus provided with the mechanism of chemical reactions. To express the above description of a reaction as

$$2HCl + Zn = ZnCl_2 + H_2$$

is a code both for the calculation of equivalent weights, and an expression of the mechanism by which the observable test-tube reaction takes place. The equation explains the reaction but at the cost of requiring us to understand, among other novelties, the new concept of *valency*.

To explain valency we need something more, given by the electronic theory of the atom which enables us to account for the valencies which we can deduce that chemical elements must have to react in the way and with the quantitative relations that are observed. It provides the mechanism for the affinities which are themselves the mechanism, or part of it, of the observed chemical reaction. We say that one atom of zinc reacts with two molecules of hydrochloric acid because zinc has a valency of 2. Furthermore, we say that zinc has a valency 2 because it has 2 electrons in its outermost electronic shell.

Chemistry then exhibits the following large-scale logical structure:

Observable reactions
explained by
atomic hypothesis and valency   (Mechanism I)
explained by
electrical theory of atomic structure   (Mechanism II)

At present we do not have a satisfactory Mechanism III, but it is in the invention or discovery of a satisfactory Mechanism III that the direction of theoretical progress lies.

EXAMPLE 2

The rest positions of magnets, their properties of attracting and repelling other magnets and magnetic substances, their capacity for magnetization and their loss of magnetic 'virtue' under certain circumstances, can all be explained by the theory of desmesnes, that is, that magnetic substances contain minute magnets, and it is in terms of this mechanism that the properties of large magnets get an elementary explanation. But we need another mechanism to explain the properties and behaviour of desmesnes. This is provided, in principle, by the electro-magnetic theory of the atom. Again we have a hierarchical structure of explanation:

Observable behaviour of magnetic substances
explained by
theory of desmesnes
explained by
electro-magnetic theory of the atom

The hierarchical relation ' . . . is explained by . . . ' is clearly asymmetrical, for if a explains b, b does not explain a. Nor is the relation reflexive since no set of statements is self-explanatory, in any sense other than 'self-evident'. It is not always clear whether the explaining is also transitive. In most cases it is theoretically transitive but practically intransitive. It is theoretically possible, for example, that the gross behaviour of substances in chemical reaction could be explained by the electronic theory of valency, indeed in some simple cases it is also a practical possibility. However, the practical possibility of such explanations is remote for reactions of any complexity and, in general, in most sciences the transitivity of explanation as a practical possibility cannot be assumed.

At each level the explanatory mechanism provides the connections for causal relations at that level. If we want to know the cause of magnetic hysteresis, we could provide a first order account by referring to the theory of desmesnes and the energy required to put them into a new configuration. It should be noticed that the explaining relation is very much more general than the causal relation, for hierarchical explanations in chemistry, for instance, do not provide the basis for what we would ordinarily call causal relations. We have in chemistry a case in which we discover production relations that do not overlap with causal relations. For instance, it is quite proper to say that hydrogen can be produced by the action of dilute sulphuric acid on zinc which contains traces of arsenic, but this is not the cause of hydrogen. This illustrates another feature of the language of causality: only events and states can be caused; substances, though produced or generated, are not caused.

In every hierarchy of explanatory mechanisms there is, at any given historical period, an ultimate or final mechanism which, in that period, does not call for an explanation. The characteristics of this mechanism, or the concepts of what counts for the time being as the ultimate explanation, can be expressed in what I shall call the *General Conceptual System* of the period.

The Nature of a General Conceptual System

The g.c.s. of any period provides, in a general form, the received standard answer to two fundamental questions, which do not call for any further enquiry so long as that g.c.s. is in favour. The two fundamental questions are:

Q. (1) What is the world made of?
Q. (2) What is the fundamental process by which changes occur?

The Aristotelian g.c.s. gives the following, quite simple seeming, answers:

A. (1) Matter and Form.
A. (2) The replacement of one Form by another.

To our two fundamental questions we must add a third which, though subsidiary, is logically necessary to a g.c.s.:

Q. (3) Given some *A*. (1) and *A*. (2) what must the stuff of the world be like for the fundamental process or processes to occur in it?

To this the Aristotelian g.c.s. gave the following answer:

A. (3) Matter is such that it has a Potentiality for Form. In the second part of this study we shall be concerned with an example, which stretches more or less over three centuries. It is the string of arguments and experiments that the philosophers and scientists of the sixteenth, seventeenth and eighteenth centuries used to establish their own g.c.s. in so far as it concerned changes in the concept of matter, in opposition to that of Aristotle.

We shall be looking in great detail at the origin and growth of a g.c.s. but it is worth noticing and bearing continually in mind that the g.c.s. of a period does not appear from nowhere as the pure invention of scientists and philosophers, but is the result of a certain selection from experience. For instance, the Aristotelian c.s. employs concepts from biology and art, but not from warfare, ship-building, mathematics, architecture or politics. Once the set of concepts for the g.c.s. has been selected then it becomes fixed in a certain abstract form, for in its concrete and original setting it is the c.s. of some particular activity or department of nature.

A general conceptual system can be expressed as a tripartite list.

(1) The list of classes of independent individuals
(2) The list of classes of properties of those individuals

These two together provide the detail of the answer to the first fundamental question above and anything mentioned in (1) and (2), or formed by some complex of items mentioned in (1) and (2), can be said to be real or to exist.

In a general way we can say that the expression 'to exist' means, in the context of a scientific enquiry anyway, 'to be a member of, or constituted out of members of, the list of entities mentioned in the accepted general conceptual system'. Trouble of a philosophical kind arises with this expression because in its ordinary or everyday use it can plausibly be identified with 'member of the class of material things'. Unfortunately that class of independent individuals is not well-defined.

(3) The list of classes of relations between individuals or properties.

This third list provides an answer to the second of the two fundamental questions. Mechanisms constructed from the listed individuals, with the listed properties, and acting and interacting in accordance with the listed relations are, for the given g.c.s., the fundamental mechanisms of nature, and close each hierarchy. The relations that are admitted between individuals and properties of those individuals are of two kinds, in any g.c.s. whatever. There are those relations which I shall call *interactions* and those which I shall call *frameworks*.

Interactions. These can either be between individuals or between properties. For instance, impact, gravity and electrostatic attraction and repulsion are interactions between individuals. Incidentally there are some puzzling cases thrown up by modern physics in which it is difficult to disentangle interactions between individuals from interactions between properties. The exchange of virtual photons is such a case. Interactions are each examples of causal relations. For instance, a moving ball impacting with a stationary ball is said to cause the second ball to move. In the interaction of properties, a permanent magnet causes a piece of unmagnetized iron to acquire magnetic properties. It is not the permanent magnet as an individual which causes the iron to acquire magnetism, but the magnetic properties of the permanent magnet causes the magnetism in the iron.

Frameworks. The kinds of relation between individuals that I am proposing to call 'frameworks' are those we use the generic terms 'space' and 'time' to describe. By means of the single relation 'temporal precedence' we can, in principle, order uniquely all individuals and properties that have existed, do or will exist. By means of the three relations 'to the left of' 'in front of' and 'above' we can completely order all individuals and properties which exist at any given time. What we cannot do with the three spatial relations is order individuals and properties uniquely, for the choice of origin and orientation for a spatial framework is arbitrary. It is perhaps worth noticing that though the single temporal relation gives a unique ordering of events (relativity has nothing to do with order), there is no unique temporal metric, for any set of event pairs can be chosen as the fundamental or standard regularity and hence as the basis for a time metric. But that the chosen series is absolutely regular cannot be known and indeed it is hardly an intelligible suggestion. To treat any series as regular is to make an arbitrary choice, though whatever choice is made need not be irrational.

The century and a half from 1570 to 1720 is commonly regarded as the period of the Scientific Revolution. So rich was this era in theories and discoveries that any account of what happened in it must necessarily be selective. Historians have tended to emphasize two particular aspects: the attainment of cosmological freedom (the break out from the Aristotelian egg) and innovation in method, in particular the development of the idea of systematic experiment and the growing use of mathematical description. These aspects of science were already present in Greek times and often dominant. Archimedes' studies are every bit as good or even better instances of the methods of mathematical physics than anything Galileo did; Aristotle's embryology is as competent an experimental investigation as Gilbert's of the magnet. The crystalline spheres only became adamantine in the Middle Ages.

There was another aspect of the Scientific Revolution which to my mind was at least as important as those which have received most recent emphasis. That was the new theory of matter and of change in material things. In discussing my example in Part 2 I want to undertake the task of bringing out this particular strand so that, though I disclaim the primary intention of an historian, my choice of the 'Corpuscularian Philosophy' as an example of a general conceptual system is neither accidental nor untendentious. It may seem surprising that Descartes finds no place in my example, but in the matter of *substance* and the substance of *matter* he initiated a tradition very different from the Corpuscularian Philosophy.

Taking up this thread discloses some new lines of connection of considerable interest. There is that from Gassendi to Newton, whose student note-books contain a précis of the *Syntagmata*. There is that from Boyle, as he apears in his *Origins of Forms and Qualities*, to Locke; then too one can spot a strand running from Bacon and Galileo to Locke in the doctrine of primary and secondary properties; and there is Newton in a new role as epistemologist. In all this one has the feeling of a new view of the way things work, neither clockwork nor geometry, but the orderly rearrangements of moving corpuscles. If we need an imaginary Hellenic progenitor it is more a dynamic Pythagoras than Democritus, since order and structure are as much key concepts of the Corpuscularian Philosophy as is the doctrine of atoms and the void.

Invariants and Principles of Conservation

In addition to the lists of individuals, properties and relations that are to be taken as basic, a general conceptual system must specify certain invariances in natural processes and properties. Without any invariant, processes or properties or quantities science, as we know it, would be impossible. In the Aristotelian conceptual system the forms were invariant and change came about by the degree to which an invariant form was actually present in matter. The Aristotelian explanation of something getting warmer was that the form of heat, an invariant, was being more perfectly realized in the matter of the object. Similarly the growth of organisms to adult form was explained as the gradual realization of the invariant form of the species, which is present only potentially in the infant.

In the Newtonian conceptual system the invariants are processes and quantities. The invariant process is the motion of a free body in a straight line, that is a body unimpeded by restraining forces. For the purposes of the invariance we do not need to decide whether the invariance of the motion defines free space or whether we take free space as given and define free motion in terms of it. Furthermore, various properties of bodies are conserved in the system, for instance it has emerged that among others there are energy, momentum and mass. To put this another way, it has emerged that there are ways of defining these concepts so that their quantity is conserved. Change, in the Newtonian g.c.s., is explained as the redistribution among objects made up of fundamental individuals, of the invariants, which are such that they can be passed from one body to another. Bodies are said to acquire energy, momentum and mass at the expense of other bodies which lose exactly equal amounts. Are there really quantities of this and that which are being transferred from one thing to another? In a way we are not obliged to answer the question in quite those terms.

These invariances are not at all obvious in nature, and cannot be demonstrated in any obvious way by experiment. It is not clear how one would demonstrate experimentally that momentum is conserved in a certain sort of interaction. This cannot be done by separating off the momentum and measuring it independently of the bodies which 'possess' it; for momentum is the product of mass and velocity and these are quite different sorts of bodily property. Clearly the origin of the invariances is not to be found in the experimental side of science. It is to be found in the general conceptual system. This is as true for Aristotle as it is for Newton. It is impossible, given the Aristotelian general conceptual system, to demonstrate experimentally its central invariance, that of form, since when it is apparently variant, it is in part not yet realized, and that which is not yet realized is beyond empirical investigation.

Why should a total flux make science impossible? Why should there not be a Heraclitean science as there has been

Aristotelian and Newtonian science? An analogy might help. Suppose that coins were made which had no particular value in terms of goods. Not only would one be able to purchase different quantities of goods for the same number of coins on different occasions, but shop-keepers would take arbitrary and different numbers of coins for the same goods on the same occasion of purchase. In these circumstances we would hardly be prepared to call this coinage a currency. It would be impossible to make predictions as to what could be bought with any given number of these coins, or to estimate what any given thing would be worth. Similarly it would make no sense to say that this was cheap and that dear, for irregularities in price would now have no meaning.

Similarly, in the sciences no predictions would be possible unless there were some invariant which 'carried through' from the initial conditions to the predicted outcome of a process. Prediction depends upon having a rule which usually works, and to have a rule a necessary condition is that it be applied successfully more than once. But for there to be a second application of a rule there must be some invariance either in process, substance or property which persists from the first application of the rule to the second and subsequent applications, as the invariance of momentum is a characteristic of mechanical changes. There must therefore be two kinds of invariance. There must be those which persist through each process considered separately, and there must be those which persist in repetition so that rules can be applied. We say 'must' here, not because there is some universal necessity which forces these invariances upon us, but because the nature of our science is determined by our choice of invariants.

One of the problems in constructing a general conceptual system is to choose relations between individuals and between properties which lead to invariances which can serve both requirements at once. To do this an invariance needs to be as general as possible, that is, it is to be exhibited in as many different processes as possible. One might say that there are different dimensions of invariance needed by the sciences. There need to be those which persist through individual processes, and these need to be in some way the same, whenever such a process occurs.

The shrewdest and perhaps the most fertile choice of fundamental relation was made by Newton. For any two material bodies in interaction the forces they exert on each other are equal and opposite. From this, by means of the definition of *force* as mass-acceleration, the invariances of momentum and, later, energy were able to be deduced. One might say that the Action and Reaction principle determines the form of all mechanical laws as the expression of an invariance in certain quite artificial physical quantities before and after a mechanical interaction. Such laws are of use to us since we have a way of eliminating the artificial quantities in favour of some more or less complex functions of observables.

A further consequence of the construction of invariants is that their use is a condition for the application of arithmetic to the description of a physical process. Since the application of arithmetic depends upon the setting up of relations of numerical equality an invariance of some quantity is clearly a necessary condition, though it is not sufficient for the acceptance of a numerical equality between certain quantities as the true form of a description of some interactions. It is also necessary that the quantity held to be invariant should be a function of quantizable observables, and should of course be a computable function. These are, as it were, the semantic and syntactic conditions of using equalities in the description of interactions. The application of arithmetic to chemistry illustrates the coming together of a number of invariances to settle the conditions of arithmetization. The Principle of the Conservation of Matter, the Laws of Multiple and Definite Proportions, and an hypothesis about the invariance of atoms, were all required before it was possible to set out an equation between the reagents and products of a reaction in terms of an equality between the equivalent weights of the elements involved. A good deal more was needed before the modern chemical equation of atoms was possible, for instance the concept of valency and the Cannizarro Memoir of 1859 distinguishing at last the ultimate units of chemical change.

An example of a third kind is that of Mendel's Laws of Inheritance. The laws which Mendel discovered depended, at least implicitly, on reorganization of constant genetic factors, invariant elements in the transmission of heredity. The Aristotelian invariance of species, or total set of characteristics which differentiated the kinds of organism, was replaced by an invariance in particular genes, that is, not in the characters of the adult organism, but in the agents responsible for the development of the characters of adult organisms. The ingenious theory of recessive and dominant genes preserved the invariance of the genetic factors while allowing wide variation in the actual characters developed by adult organisms. This opened up quite new conceptual possibilities though these were not grasped immediately. If species are the sum of genetic possibilities then, since it is the genes which are invariant, species could, by genetic shuffling, be changed. Here, though the invariance condition for arithmetization had been satisfied by the invariance of species theory, the 'species' was not an easy notion to quantify. Mendel's genetic factors, on the other hand, could be counted and their proportion in a population determined. So, with the invariance of these factors, the two conditions for arithmetization were satisfied. Of course Mendel barely scratched the surface of the network of genetic laws but no conceptual innovation was needed to allow the modern molecular code theory of Watson and Crick to be formulated. This is as much Mendelian biology as the work of Herz and Einstein is mechanics, in the style of Newton though, rather ungraciously, we contrast the latter with Newtonian mechanics.

Modifications to a General Conceptual System

Two kinds of modifications are possible to a g.c.s. I shall call these 'scientific' and 'philosophical' modifications respectively.

To make a scientific modification to a g.c.s. is to change the items in one or more of the three lists. We might, for instance, decide to say that chemical atoms are not the basic individuals, but that some other entities are, e.g. electrons, protons and neutrons. We might, and indeed have, decided that it is no longer profitable to treat species as the basic individuals subject to evolutionary change, and to substitute for these basic individuals new entities, populations. To make choices of this kind is to modify our g.c.s. by changing the content of the list of existing individuals. However, we cannot change the items in a list independently of items in other lists. If we choose to drop atoms in favour of electrons, protons etc. then we must add electro-magnetic properties to our list of fundamental properties, and electro-magnetic interactions to our list of fundamental relations. Similarly, to choose impenetrability and extension as the fundamental properties of individuals and to exclude the Aristotelian notion of form, forces us to adopt a different set of basic individuals and to modify our list of interaction relations from the mutual replacement of forms, to the impact between particles.

To make a philosophical modification to a g.c.s. is to claim to reduce the items in one or more lists to the items, or some combination of the items, in another. For instance, the phenomenalist doctrine that things are congeries of sense-data is an attempt to force the reduction of the items in the list of basic individuals to combinations, or functions, of the items in the list of basic properties. To claim, as Descartes does, that the

properties individuals seem to have are the result of the interaction of organisms and things, is to claim to reduce at least
part of the list of basic properties to some combination of basic
relations. Hobbes, for instance, makes the same move in suggesting that 'outer sense' is the reaction to the impact of atoms
on the brain. Again, to claim that the physical properties of
things are really the observable consequences of different con-
figurations of their atoms is to claim that some items in the
list of properties are to be reduced to functions of items in the
lists of individuals and relations.

If a general conceptual system requires, or is thought to
require, modification, this can be achieved either by the sub-
stitution of new items for old and the consequent modification
of all the lists of the system, or by the reduction of items in one
or more lists to items or combinations of items in the others.
In those scientific revolutions which have actually taken
place, we find that not only is one general conceptual system
reached from another by scientific modification, but that
there are consequential philosophical modifications too. For
instance, the Newtonian system of concepts as enunciated by
Boyle shares with the Aristotelian 'one catholick matter' but
its individuals are not constituted by the realization of Form
in Matter, they are independently existing corpuscles. The
only real properties of corpuscular matter are 'bulk, figure,
motion and texture', so other apparent properties of matter
must be shown to be reducible, in some way, to those. Thus,
Universal Matter disappears. To argue for the corpuscularian
nature of reality is to argue for a scientific modification to the
general conceptual system. To argue for the distinction between primary and secondary properties is to argue for a
philosophical modification to that conceptual system which
is undergoing change. However, before we investigate more
closely those pressures which tend to change general conceptual systems we need to understand in more detail the function
of a g.c.s. in science.

The Function of a General Conceptual System

Passing reference has already been made to the function of the g.c.s. in providing the materials for 'ultimate explanations'. We have already seen that this is not the only function of the conceptual system. It also provides the materials for standard or acceptable reductions of observational concepts, those I have called philosophical modifications. The general conceptual system not only provides the mechanism for explaining, for instance, the observed relations between calorific phenomena, but at the same time provides the materials for acceptable claims as to what calorific phenomena 'really' are. Indeed, that some sort of reduction of observational concepts to concepts of the g.c.s. can be made is a necessary condition for the way in which descriptions of g.c.s. mechanisms can serve as ultimate explanations. If our general conceptual system is 'Atoms in Motion', then all calorific phenomena are to be understood as either the observable effects of atoms in motion, or atoms in motion observed from the point of view of a human observer or one of his instruments. In the case of the modal link between g.c.s. phenomena and observable phenomena the necessity for reduction of observational concepts to g.c.s. concepts is obvious, for that is just what a modal link is. Using the general conceptual system, we can say that what we observe as temperature is, in nature, the kinetic energy of the molecules of which substances are 'really' made up. So that for scientific purposes there are not two independent properties of substances, their temperature and the average kinetic energy of their molecules, but only one property which is observed as temperature. In the case of causal links between theoretical mechanism and observed phenomena, the observed (effect) phenomenon must be modally linked to a g.c.s. property or phenomenon. For instance, it is no good saying that gas pressure is the effect of the changes of momentum of gas molecules at the phase boundary unless we have reduced, in the modal sense, gas pressure to the sum of the mechanical forces exerted by the molecules upon the boundary surface of the gas container.

Similarly, for the current theory of inheritance to be convincing, not only do we need to know the structure of the molecules of the genetic material, but we need to be convinced that the observable characteristics of organisms are, in fact, modally linked to chemical structures. We need to believe that what we see as the blueness of eyes, for instance, is, in terms of our current g.c.s., a certain molecular and electronic structure which separates certain wavelengths of light. It is this molecular structure which is causally linked to the structure of the D.N.A. in the nuclei of the cells of those organisms having blue eyes. The general conceptual system functions in a dual capacity, providing the material for the explanatory mechanisms of the sciences, as well as providing the materials for the reduction of observable phenomena to a form which could be causally linked to the explanatory mechanisms.

The forces which make for change in a general conceptual system can come from sources linked with both these functions. It may be that the g.c.s. fails to provide a convincing and unified account of the mechanisms by which observable phenomena are produced or generated, or it may be that phenomena will turn up which prove intractable to the form of reduction determined by the g.c.s.

Conceptual Systems and Reasoning

(A) REGULATIVE PRINCIPLES

To derive his concept of inertia, Newton had implicitly to suppose that the mass of a body was not affected by its location,

D

chemical properties, or by its undergoing acceleration. Why this was so Newton was not obliged to say, since it was in terms of this invariance that everything else in his universe was explained. In addition to the invariance of mass, Newton makes inertial motion a fundamental process and hence one not requiring explanation. There is no need to·explain why a body continues in its state of rest or uniform rectilinear motion; in the Newtonian system it is logically impossible to supply such an explanation. Change of place is a process defined as uniform within the system, that is, no change in it will occur without action from outside the mover, and hence is not itself open to explanation. What must be explained is change of rate of change of place, and to these explanations forces are perfectly adapted since they are defined as the causal agents for just such processes. Since forces are the only causal agents in Newtonian mechanics, where forces do not act we do not have the material for a causal account of what is happening. Hence, inertial motion and rest are, for Newton's system, necessarily the fundamental and unexplained processes. If there is no reason why a change should occur (where 'reason' is defined within the system as the action of its allowable causes) then no change does occur. No science is, therefore, able to account for everything that happens; not just as a matter of empirical fact, but as a matter of logic. Somewhere in a science there must be a process for which there is no reason why it should run thus, and so, and not in some other way. By 'no reason' here is meant 'no statement to the effect that a member of the set of admissible causes acting'. Newton does not need, and hence constructs his system so that he cannot supply, an explanation of inertial motion. In the Aristotelian system, on the other hand, no reason can be given in its physics for downward appetition, but having chosen to make that process fundamental, what we now call inertial motion would require explanation. By choosing to make the following identifications:

(*a*) The only caused changes are accelerations and decelerations

(*b*) The only causal agents are forces

Newton closes one channel of enquiry by opening others.

In general we could say that if we observe or imagine *n* different types of process occurring, only *n* – 1 of those types can get an explanation in any given scientific system. The qualification 'or imagine' is necessary, since, as we have already seen, explanation proceeds most frequently from theoretical entities, events and processes to observed entities, events and processes.

We get from this, two regulative principles closely connected with the application of the general conceptual system.

The Principle of Sufficient Reason. For any change which we can observe or detect, a sufficient reason to explain the nature and direction of such change is to be found in the .entities, properties and interactions of the general conceptual system.

This leaves the processes mentioned in the general conceptual system unexplained, but not for ever inexplicable. It is always possible that a new general conceptual system should be adopted in terms of which the previous g.c.s. becomes explicable. We might express this possibility in another principle.

The Principle of Universal Causality. For every natural process or change, a causal account can in principle be supplied.

This principle is operative, in particular, at those times when conceptual change of a drastic kind seems to be called for. The Principle of Sufficient Reason is limited in application to all phenomena other than those mentioned in the lists of the g.c.s., whereas the Principle of Universal Causality, applying as it does to all phenomena, could be understood as expressing the methodological hypothesis (rather than principle) that for any conceptual system which might be adopted another could be constructed which would stand in an explaining relation to the processes and events describable by the first.

(B) COGENCY AND THE GENERAL CONCEPTUAL SYSTEM

All forms of deductive reasoning, that is of drawing conclusions from given premises, can be expressed in such a way that each step is governed by the same logical principle *modus ponens*. This principle can be expressed in several ways, but for my purposes I shall take the following as the standard form:

If '*p*' is a premise and '*q*' the conclusion one draws from '*p*' then the argument '*p* therefore *q*' is sound or cogent, in so far as the conditional proposition 'If *p* then *q*' is acceptable.

In formal logic the conditional proposition 'If *p* then *q*' is analysed entirely in terms of the truth-relations which hold between the truth and falsity of its components '*p*' and '*q*': '*p*' and '*q*' are considered quite independently of each other, and on connection between them is supposed and the consequent truth or falsity of the compound conditional statement formed from them. This analysis is commonly expressed in a table as follows:

p	*q*	if *p* then *q*
True	True	True
True	False	False
False	True	True
False	False	True

The truth-relation between statements which is defined by this table is represented by the formula '$p \supset q$'. Clearly this truth-relation does not depend at all upon the particular meaning which '*p*' and '*q*' may have in actual arguments. What the table expresses is a necessary, though not a sufficient, condition for the cogency of an argument from a premise, or premises, represented by '*p*' to a conclusion represented by '*q*'. Another way of putting this is to say that what the table expresses and hence what the truth-relation '$p \supset q$' expresses is the condition that an argument in which the premise, or premises, are as a matter of fact true and the conclusion as a matter of fact false, is not a valid or sound argument, no matter for what statements '*p*' and '*q*' may stand.

To find what are the sufficient conditions for cogency of an argument we need to examine other relations that might hold between premises and conclusions. Roughly we can say that '*q*' follows from '*p*' if we know that some general connection exists between whatever '*p*' stands for and whatever '*q*' stands for. For instance, if we know that what '*p*' describes is the cause of what '*q*' describes, then, provided that the assertion that *p* is not true when the assertion that *q* is false, the argument from *p* to *q* is cogent. Two kinds of relation must hold between premises and conclusions for arguments to be cogent in the strongest sense. There are those which depend upon what statements are actually being made in premises and conclusion, that is depend upon the meaning and not just the truth of the components of the argument. Then there are the formal truth-relations. Among those which depend on the content of the clauses of an argument there may, for instance, be causal relations, where '*p*' and '*q*' stand for assertions that such and such events, states or processes have occurred. There may be identity relations where '*q*' simply states or describes some different aspect of '*p*' and so on.

Causal Relations

There is rain in the hills
therefore
The plains will flood

Two conditions must be satisfied if this argument is to be cogent.

(*a*) It should not be the case when it is true on a particular occasion that there is rain in the hills the plains do not flood.

(*b*) It should be generally true as a geographical and meteorological fact that rain in the hills is followed by floods in the plains.

That both relations, the particular truth-relations and the general causal relation, are required is demonstrable by considering the conditions for the defeat of this argument. If it should, as a matter of fact, turn out to be the case that there is rain in the hills and no floods, then the argument is unsound, even though this may not be counted as an instance counter to the general geographical truth. Someone may have built a dam. The truth-relations are to some extent independent of the general statement. But if we should know as a general truth that there is no causal connection between rain in the hills and floods in the plains, then whatever the truth-relations of premises and conclusion turned out to be on a particular day, the argument would not be cogent, even though the minimum conditions for validity had been satisfied.

Identity Relations

This is anthracite
therefore
This contains carbon

Again we find that two conditions must be satisfied before we can regard the argument as cogent.

(a) It must not be the case that as a matter of fact this is anthracite but does not contain carbon.

(b) Anthracite is a form of carbon.

In this case the failure of the truth-relations in a particular case would be a counter instance to the general identity statement, for we do not admit the logical possibility that genuinely non-identical things can be identical to a third thing (this would be a test for concealed non-identity). Contrast this with the previous case of causality. There we do have the genuine logical possibility that genuinely different causes should issue in the same effect.

Analytic Relations

This is gold
therefore
This has a specific gravity of 19·6

Provided it is true that this is gold and not at the same time false that this sample has the required specific gravity, and provided the current received definition of gold includes the clause that it has this specific gravity, then the argument is sound. Again the failure of the truth-relation leads to yet another situation. We have the option of denying that the specimen under investigation is gold, and putting the weight of identification on the specific gravity; or of removing specific gravity from the definition of gold, putting the weight now on those other properties which are used merely in the identification of this element.

The general conceptual system is not directly involved in any of the argument forms above. It would not be right though to suppose that it follows from this that such arguments can be cogent independent of the general conceptual system. Of course they can be valid independently of any subject matter whatever and hence of any general conceptual system. Putting this another way, one might say that the best way to discover what one's g.c.s. is, is to notice just what gap there is between validity and cogency in any argument. The cogency of a piece of reasoning, depending as it does upon causal, identity and meaning relations, is dependent upon whatever these relations are dependent upon. These, being matters of fact or closely connected with matters of fact, depend upon what way the world is being viewed, and thus depend upon the general conceptual system. We might say the general conceptual system underwrites all such particular factual statements in a certain way.

To get at this relation of 'underwriting' we need to look into the logic of the relation between factual statements and those statements which express whatever sort of world view the factual statements presuppose, for the general conceptual system consists of lists of concepts expressing the ultimate assumptions of existence, process and structure of the world.

The elucidation of the general form of the presupposition

relation we owe to P. F. Strawson.* Suppose we have a statement 'p' in which reference is made to some individual or class of individuals 'i', property or class of properties 'j'. Then it is said that the following relations hold:

'i or j exists'	'p'
True	True or False (Open to test)
False	Pointless

Applying this to the involvement of the general conceptual system in reasoning we can see that, provided reference is made to entities, properties or processes mentioned in the lists of fundamental items, statements are automatically guaranteed not pointless, for the general conceptual system lays down what there is, and thus such statements are open to test of truth or falsity, and arguments employing them to tests of cogency. However, arguments which depend upon alleged factual connections which contain reference to items in an outmoded or unacceptable general conceptual system are just as pointless as are the general connections allegedly guaranteeing them; whether they satisfy tests of formal validity or whether they do not. For example:

> This substance leaves little ash when burnt
> therefore
> This substance is rich in phlogiston

is an argument that is at best an historical curiosity and can play no part in modern chemistry, not just because the particular conceptual system of modern chemistry does not contain the concept 'phlogiston', but because the *general* conceptual system of modern science does not sanction the concept 'material fluid of negative weight'.

As in this example, on many occasions the reference in the general factual relation guaranteeing an argument is not directly to items in the g.c.s., and there may be more or less

* Strawson, *Introduction to Logical Theory:* Chap. 6, Pt. iii, Sect. 7.

extended series of implications from the actual referent to an item in the g.c.s. But provided that after a finite number of steps an item or items in the g.c.s. is reached the argument is not pointless, though it may be both invalid and not cogent. In a similar way the general conceptual system underwrites identity relations. For instance, if an argument depends upon the proposition 'Temperature is the observable concomitant of the average kinetic energy of the molecules of a gas sample', then since this statement is dependent for its having point upon there being such items as molecules, so is the original argument. Molecules do not figure in the Newtonian general conceptual system, but we have the further proposition that molecules are groups of atoms, and since atoms do figure in the Newtonian g.c.s., the proposition 'Temperature is the observable concomitant of the average kinetic energy of the molecules of a gas sample' is not pointless in Newtonian science. Finally, the general conceptual system is involved even in meaning relations. To define elements by their atomic number presupposes not only that there are atoms but that they are built up in a certain way out of certain items. So, though 'Gold has an atomic number N' is not meaningful in Newtonian science, it is for us, since our general conceptual system contains those items the assumption of whose existence is required for the intelligibility of the statement.

The role of the general conceptual system is thus to give particular form to the regulative principles needed for any form of science, and to underwrite, by settling in advance, questions of existential reference, one of the necessary conditions for cogent reasoning in the sciences for which it is the conceptual system.

The General Conceptual System and Existence

A classical philosophical issue must now be faced. It is the general problem of how we are to understand assertions of

existence. We have found that we want to be able to say that
the kinds of things which exist are specified in the lists of the
general conceptual system. To anticipate any difficulties there
may be in this answer for some readers—there is a suggestion
of conventionalism here that may be unwelcome to some—
it will be necessary briefly to sketch out a system of existential
concepts slightly more elaborate than that usually deployed
by philosophers and others. The elaboration of the system of
existential concepts is forced upon us by the two basic demands put upon philosophers by the standards of their trade,
consistency and adequacy to the requirements of all that must
be said. What I shall have to say is closely connected with my
analysis of existence claims in *Theories and Things* (Newman
Series No. 9), but the problem is approached from a different
direction.

REALITY AND EXISTENCE

From Kant we learn that 'existence' is not a predicate in the
logical sense, that is, is not the name of a property which things
may have or lack.* From Moore we have the suggestion that
since 'existence' is a predicate in the grammatical sense it must
have some role in language and that its sense is rendered by
the equivalent quantifier 'There is a . . .'† From Lesniewski
we learn that paradoxes cannot be avoided by this device and
that a formally adequate concept can be devised with the help
of the rather surprising suggestion that we treat the statement
'*A* exists' as equivalent to '*A* is a member of some classes but
not of all'.‡ Formally satisfactory as the latter proposal is, it is
too far removed from our usual ways of expressing ourselves to
be helpful in this present context. I shall therefore outline a
related system of existential concepts which is, I hope, both
consistent and adequate to our needs.

* Kant, *Critique of Pure Reason:* Transcendental Dialectic; Bk. II, Chap. iii, Sect. 4.
† Moore, *Logic and Language*, II, Chap. v.
‡ *British Journal for the Philosophy of Science*, Vol. 5, p. 104.

We shall need two ranges of existential concepts. There are those of which 'real' is a typical example, and there are those of which 'exists' is a typical example. We can see their uses in some examples. Suppose we find a golfer striking furiously but ineffectually at a round, white patch of sunlight on the grass under the trees. We might elucidate the situation by pointing out that the apparent golf-ball was not real. This would not be to deny that a patch of light existed on the grass though. It would be to point out an important distinction in what appeared to be round, white and lying on the ground. This contrast use of 'real' may fall across the world almost anywhere. But since one of the ways of being unsatisfactory is just not to be at all, sometimes the real is the existent and the unreal just limbo.

I believe that we can make the best use of 'exists' itself by taking up something of Lesniewski's idea and building a set of concepts on that basis, one of which turns out to be among the standard uses of 'exists'. I construct my set of existential concepts as follows: we treat the statement 'As exist' as asserting that a class constructed in a certain way has members. The class is not just the class of As, for that would not distinguish the unicorn from the okapi, but is the class formed by the intersection of the class of As with some standard class implicit in the discussion. For example, if the implicit class is 'contemporary African mammals' then the unicorn is correctly said to be non-existent since the class of one-horned horses which are contemporary African mammals is empty. On the other hand the intersection class of camel-like equine quadrupeds and contemporary African mammals is not empty—okapis exist. But if the implicit class had been European mythical beasts then unicorns do exist in that context, while the Taniwha does not. From this it follows that the suggestion that it might have been a Taniwha and not a dolphin which carried out those spectacular rescues in the ancient Aegean is absurd. I shall call the classes implicit to discourses of various kinds *ontological classes*; and those classes whose actual having or lacking members is what is specifically asserted in an existential statement I shall call *ontic classes*, and I shall say that the specific classes which form one of the components of the ontic classes, are *descriptive classes*.

Now it can be seen just how the general conceptual system would function in this scheme. It would lay down, before any approach is made to the world, the ontological classes. Of course we would choose those ontological classes which we had reason to believe would have the widest application. Empirical science, having suggested the outline of our approach to nature, would do the rest. The possibilities of observation and the capacity of instruments determining the descriptive classes allows the question of the membership of the ontic classes formed by their intersection to be settled by empirical means. To put this colloquially we could say that the general conceptual system we have adopted determines the kinds of things, properties and processes we are prepared to admit but what there actually is will be found out by investigations such as turning over stones and analysing the electrical impulses in the circuits of radio telescopes.

A good many of the apparent difficulties involved in giving answers to existential questions derive from the way we carry round with us a particular ontological class—that of things as perceived, often modified towards the material object of Newtonian science. This has been true of recent advances in physics where the oddness relative both to perceived things and to Newtonian material objects of the entities stumbled upon or postulated in accordance with the demands of phenomena, has led to quite ridiculous rulings out of existence that would take the whole world with them.

EXISTENTIAL CONTRASTS

The contrast between the real and the unreal is not necessarily identical in all contexts with that between the existent and the non-existent. This we have already remarked. Further refinements are needed. When we declare that A is really B there are at least two distinct senses in which this adverb might be used. We may want to draw attention to the fact that A is analysable as B or we may want to draw attention to the fact that A is to be replaced by B, that to call what we have now chosen to call B, A, is a mistake. In either case the legitimacy of B is dependent on its relation to the items in the general conceptual system, but in the former case A exists as much as B, while in the latter case A does not exist but is replaced by B.

Here are some examples from the sciences of these contrasting uses of 'really is'. If someone says 'That salt crystal really is a cubical lattice of sodium and chlorine ions' he is not intending to deny the existence of the salt crystal. In each case the ontological class of both analysans and analysandum is the same. On the other hand if it is said that flying saucers really are reflections on clouds there is the clear intention of denying the existence of flying saucers, and here the implicit ontological classes are different. On the one hand there is the class 'material object' and in the process of reduction the phenomena of the ontic class are reinterpreted and assigned to a different ontological class, namely reflections.

Whether or not an analysis or a reduction is correct will depend upon the outcome of practical investigations, but whether or not an analysis or a reduction is legitimate will depend upon whether the preferred ontological class is sanctioned by the general conceptual system which has been accepted. This latter requirement has point, particularly in the construction of theories where all we may have to go on in existential decisions is the legitimacy of the concepts.

We have noticed that an accepted general conceptual system underwrites analyses and reductions. Looked at in another way this relation between analyses and reductions, ontological classes and the general conceptual system can also provide a route for the establishment of the latter. The power of certain ontological classes like that of the Newtonian material object in providing the basis for analyses and reductions is as good support as one could get for the Newtonian general conceptual system.

The Newtonian Conceptual System

The easiest way of devising a model for physical reality, of choosing something upon which to model the entities referred to by explanatory theory, is to make some selection from the world that we perceive. The problem is always that of deciding just what selection to make from among the many properties which we see perceived things to have and perceived phenomena to exhibit. It is necessary to choose from the multiform world a uniform set of items most suitable for the explanatory job.

The revolution of the seventeenth century consisted in choosing for physical reality, things from among the various items in the world we perceive, and in choosing from among the many properties which things may have only their bulk, figure, motion and texture and declaring these to be real. This is not just to mark off primary from secondary qualities, but it is also to distinguish tertiary qualities, as Locke called them, the powers in things that make them act as they do. There are other properties which things have—a large class of things are the organisms which have such properties as irritability and show the distressing habit of ceasing to exhibit characteristic properties when finely enough divided up. The choice of bulk, figure, motion and texture must not be supposed to have been so obvious as it looks to us now. Indeed one might say that much of subsequent speculative metaphysics was an attempt to draw up other models for reality by making different choices out of the properties of the perceptual world than those required for Newtonian science.

THE STRUCTURE OF THE PERCEPTUAL WORLD

The ordinary world in which we take ourselves to be living is made up of things, and these things are separated in space and temporally enduring. But the simplicity of this is deceptive. Psychological investigations have shown, what was often anticipated by philosophers, that perceiving is a joint process of sensing and mental integrating of sensations, often with involvement of past experience. Our perceptions of things are the outcome of the joint process. The integration of sensations covers all our sense modalities, but is, most importantly, integration of touch and sight. One way of analysing the perceptual world is to try to separate out the sensations which are one of the ingredients in our perceptions, and to try to discover by this means which sense-modality is indispensable. This may seem remote from the kind of analyses made by scientists, but its relevance will become plain.

To determine the order of priority of the sense-modalities consider these two cases:

(i) Suppose someone sees what appears to be a thing on the side of a piece of glass distant from him. He passes his hand behind the glass, but encounters no resistance. He concludes that there is nothing there.

(ii) Trying to pass our hand through invisible glass teaches us that it is a thing.

From these cases and others like them we are inclined to conclude that touch is the prior sense, in the perception of things. This conclusion can be expressed in two practical rules:

R1. If x can be touched, then whether or not x can be seen, x is a thing.

R2. If x cannot be touched, then whether or not x can be seen, x is not a thing.

(The efficacy of these *practical* rules is partly due to the rarity of tactile illusions.) The trouble which the Greeks had over deciding the nature of the stars illustrates the *thing* ambiguity which arises in cases where we are confined to sight alone.

Other *thing* doubts troubled the Greeks, for instance whether air was a substance. This doubt illustrates something else about the basic structure of the perceptual world, for the wind can be felt. The doubts arise because the wind is penetrable. It seems that of all the specific sensations of the sense of touch resistance and lack of resistance are most fundamental to our concept of *thing*.

The basic structure of the perceptual world would then seem to be regions of relative impenetrability separated by regions of relative penetrability. In the *perceptual* world we have no reason to suppose that absolute impenetrability and absolute penetrability actually occur. To christen the regions of relative impenetrability 'atoms' and the regions of relative penetrability 'the void', is to push the differentiation of regions which is relative in the perceptual world, to an absolute differentiation which is not given in perception, but which for all that may be essential to science.

The secondary structure of the perceptual world arises from the very large measure of correlation which exists in our perceptions between the sensations in other sense-modalities and the regions of impenetrability given by the sense of touch. On the whole, where we can feel something we can also see something, and we soon learn to say that we hear, taste, and even smell something there too. It is now very easy to go on to say in answer to the question 'What is a thing?' '*A thing is a complex of coexisting sensations at a region of relative impenetrability*', which is itself defined, for instance by Locke, in terms of the sensations of relative resistance. Whether or not this is a proper answer we shall discover in Part 2, but it is most important to grasp that it is a most tempting answer.

However, whatever analysis of perceived things we adopt there are some other features of the perceptual world which it is important to describe tidily. Between things there are certain relations. Things are related left and right, before and behind, above and below. And these relations can change, and in those changes we recognize the corresponding movements to and fro, fore and aft, up and down. These relations, or if we think in terms of movement, these degrees of freedom, *are* space. Any given set of relations among a set of things I shall call an *arrangement*; and a change in the relations, that is one set of relations being replaced by another, a *rearrangement*. One form of simple change is rearrangement.

But we know that it can also happen that while the spatial relations between things remain the same, the original properties of the things can be replaced by other properties. In each sense-modality we can recognize change of this sort; soft to hard, bright to dark, hot to cold, red to blue. Any given set of determinate properties at a region of relative impenetrability I shall call a *constitution* of a thing, and the replacement of any determinate property by another, a *reconstitution* of a thing. One form of simple change is reconstitution.

Rearrangement and reconstitution is complex change. 'Time' is the name we give to successive changes, either simple or complex. The passing of time is nothing but rearrangement or reconstitution or both.

This is the perceptual world. And we have already landed ourselves with one philosophical problem and the beginnings of another. Should we identify the properties of things with the sensations which we know to be ingredients of perceptions? Must this analysis of the perceptual world be the basis of all pictures of nature?

NATURAL REGULARITIES

The perceptual world not only exhibits changes, but there are certain changes which are regular. We recognize two classes of regular change, regularities of coexistence and regularities of succession. We observe arrangements of things and properties occurring together, and we observe arrangements of things or properties succeeding each other.

E

I shall say that there is a *regularity of coexistence* when after each n rearrangements or reconstitutions $(n \geqslant 1)$ from some arrangement S or constitution C, the $n + 1$ th arrangement S' or constitution C' is sufficiently similar to S or C.

I shall say that there is a *regularity of succession* when for some k, if the kth arrangement is S_1, or the kth constitution is C_1, the $k + 1$ th arrangement is S_2, or the $k + 1$ th constitution is C_2; and whenever S_1' is the nth arrangement or C_1' the nth constitution, the $n + 1$th arrangement is S_2' or the $n + 1$ th constitution is C_2'; where S_1 and S_1', S_2 and S_2', C_1 and C_1', C_2 and C_2' are sufficiently similar.

Recognition of regularity turns upon the recognition of arrangements and constitutions as *sufficiently similar*. Clearly such recognition depends upon a comparison between S_1 and S_1' etc. But if S_1 and S_1' are temporally separated then S_1 does not coexist with S_1', and this is true for each pair of arrangements or constitutions whose similarities we want to compare. Primitively we depend upon remembering what S_1, S_2, C_1, C_2 were like and using our memory in making the comparison. The unreliability of this primitive method of the recognition of regularity is replaced for science by the construction of records of S_1, S_2, C_1, C_2, which will coexist with S_1', S_2', C_1', C_2' so that a kind of direct comparison is possible. Star charts and land surveys are early examples of such record keeping. From pictures of arrangements and constitutions records develop into true notations, where certain standard symbols represent certain arrangements and constitutions.

This structure of regularities is clearly recognizable in the perceptual world. We do find arrangements and constitutions repeating themselves, e.g. the relative positions of the planets; and we find regular successions of arrangements and constitutions, e.g. the growth of plants from seed to harvest. This simple structure of regularity may need elaboration as we proceed but it will serve as the basis for general discussion.

The scientific description of nature consists of general statements describing the regularities recognized at any one time;

and we begin with the regularities of the perceptual world. With the statements of regularities go ancillary statements in which are described the exceptions to and limitations upon the regularities that have so far been recognized.

THE NEWTONIAN GENERAL CONCEPTUAL SYSTEM

From the above analysis of the perceptual world the Newtonian system is made up as follows:

(1) *Basic Individuals*
   Atoms: minimal divisions of matter
(2) *Basic Properties*
   Mass (quantity of matter), extension (quantity of space occupied), figure or shape
(3) *Basic Relations*
   (*a*) Frameworks: spatial and temporal relations—motion being the rearrangement of atoms in space during time
   (*b*) Interactions:
      (i) Impact: the impingement of one atom or body of atoms on another, the subsequent motions determined by the relations expressed in the laws of momentum and elasticity.
      (ii) Gravity: the action of one atom or body of atoms on another, at a distance, according to the law of inverse squares.

The Newtonian system of laws is constructed out of these materials by the theoretical concept *Force*. As a matter of historical fact it was not recognized by the builders of this g.c.s., Galileo, Bacon, Boyle, Newton and Locke, that force was not a member of the general conceptual system. It was not until Mach demonstrated its theoretical character that it became clear that the Newtonian *g.c.s.* did not necessarily have to include it, though for logical reasons the Newtonian *system of laws* must include force or some other theoretical concept which will reticulate the facts of dynamics.

The aim of this present study can now be more exactly expressed. It is to examine the arguments and reasons put forward by the founders of this g.c.s. for the selection and inclusion of just those items which are included in the lists, and to distinguish the scientific revolution of the period from the complementary and connected philosophical revolution. Thus, we must distinguish the arguments and reasons put forward on behalf of the Corpuscularian Philosophy which involved, in my terminology, scientific modifications to the current g.c.s. from the consequent need to argue for, and give reasons in support of, the philosophical modification of that g.c.s. which we know as the doctrine of primary and secondary qualities.

However, the purpose of this study is not strictly historical. The examination of these arguments and reasons in their historical setting is intended to bring out the logical form of the arguments and the cogency of the reasons, in the hope that some sort of general picture of the nature of conceptual change in science can be constructed.

Part 2

THE CORPUSCULARIAN PHILOSOPHY

Perceptions and Things

Perceived things have extension, solidity, weight, warmth, colour, shape, taste, smell and sonic qualities, not to mention those properties which only instruments can detect. The problem for the constructors of any general conceptual system is to decide which, if any, of the qualities perceived in objects or the properties detected by instruments make up the 'thinghood' of the basic individuals of the general conceptual system, what are in fact the fundamental and real properties of the material world. Should we adopt all the qualities of the objects we perceive or should we make some selection from these, or should we drop perceived qualities altogether from our general conceptual system and rely only on what our instruments detect?

As far as what we perceive goes, all the qualities of perceived objects are on a par. There is nothing in what we perceive which could be used to distinguish between qualities which the natural world really has, that is qualities which are also properties of matter, and qualities which, though perceived, are not properties of matter, but qualities of our percepts only. For instance, anything we perceive to be extended and to be of a certain shape is also coloured. Which is the wholly perceptual quality and which is the 'real' property? The colour or the shape, both or neither?

There are two extreme positions we might adopt when faced with the question of how far our percepts resemble the actual things in the world. We could say that they resemble them absolutely, that all perceived qualities are real properties, that there is only one kind of thing in the real world, the perceived thing. Phenomenalism is a modern form of this doctrine, though it was also held in a different form by Berkeley in the eighteenth century. On the other hand we might say that none of the perceived qualities are real properties of the real elements of the architecture of the world. For that reason we would then say that we could have no idea at all of what the real elements of the world are or what they are like; we can talk only about our perceptions. This is, in essence, the view worked out in great detail by Kant. In both cases we are advised not to make a practical distinction between what we perceive and what there really is, in the first case because we are assured that there is nothing else but what we perceive, and in the second case because we can know nothing whatever about anything other than what we perceive, except, it is claimed, simply that it exists. Neither of these positions was, it seems to me, held in antiquity.

The Aristotelian conceptual system which the Newtonian g.c.s. replaced had this similarity with its rival: that not all perceptual qualities were of the same existential status. Some were singled out as more real than others. It was not, however, that in the Aristotelian system there were hidden properties which generated the qualities of the things we see. Motion, colour, shape, temperature and so on were all forms for which prime matter had a potentiality. Each quality could be present to a greater or lesser degree. But the perceptual qualities were sorted into spectra between polar opposites; only one extreme of which was real. The polar opposite of some real quality, though perceived as a positive quality, was held, in reality, to be only the privation of the real quality. Black, though perceived as a positive quality, is, in reality, the privation of colour; cold, though perceived as a positive quality, is, in reality, the privation of heat, and so on. So far as I can see the choice of what pole to regard as real and what to regard as a privation is arbitrary.

Plato's doctrine of Forms has some obvious affinities with Kant's system, since a sharp distinction is drawn in the Platonic g.c.s. between what can be perceived and what is real.

Such a distinction is made, for instance, in the parable of the cave, in which perception is suggested to be of nothing but the shadows of reality. But, in contrast to Kant in some moods, Plato held that the thing in itself, the reality behind the appearance, could be known, but not by any kind of improved perception. The route to real knowledge was, one might say, by logical analysis. Particular, fleeting facts were not worth knowing by comparison with the necessary, and hence eternal conditions for our·having the kind of experience we do.

The philosophers and scientists of the seventeenth and eighteenth centuries, by making an 'Aristotelian' distinction among perceived qualities, and then by making a 'Platonic' use of that distinction arrived at something radically new— the distinction between *primary* and *secondary* qualities. Primary qualities are those which are both perceived qualities and also properties of actual things. Secondary qualities, though they are perceived as single positive qualities, are not matched by a corresponding singularity of secondary properties, for secondary properties are products, resultants or compounds of primary properties. We shall be concerned with this distinction in some detail, for once it was established that there could be such items as atoms, this distinction was used to give substance to the ultimate corpuscles. Thus the Newtonian general conceptual system was established. Out of the multiplicity of qualities which are perceived just four—bulk, figure, texture and motion—survived to become the real properties of matter.

The Corpuscularian Philosophy

GASSENDI

For the advocates of the new science of the seventeenth century the problem seemed quite simple. How is one to think of the world and the changes that occur in it? As the sliding in and out of actuality of fixed forms, informing an ultimate and un-differentiated matter as the Aristotelians would view it, or as the visible and tangible effects of the dance of the atoms, whose combinations and recombinations are the ultimate changes? The alternative answers are easy to state in this vague and schematic form. To confront the Aristotelian general conceptual system with another, based upon the atoms, but as powerful and as satisfying, was a task of the greatest difficulty. This was partly because it was not always clear where philosophical argument and where empirical fact was suitable to carry away any particular fortification of the dying system.

Two different kinds of question naturally come together here. There are straightforward empirical questions which are such that affirmative answers to them would support a corpuscularian view but negative or doubtful answers would have no particular result in terms of support for some view of the ultimate nature of matter. Consider, for instance, the question whether motion can yield qualitative change. If we set the parts of a body vibrating and the body changes qualitatively, say becomes warmer or redder, then we have good grounds for supposing that the quality consequently present is the outward and visible manifestation of internal motions in the parts of the body. Consider also the question of whether simple division can yield qualitative change. If we divide up a body into smaller and smaller pieces and it then exhibits a qualitative change, as, for instance, glass changes from being transparent to being opaque upon division, then this can be taken as grounds for the conclusion that the new quality, opacity, is a function of the divided state of the material. It is not difficult to conclude from these considerations that if warmth is a function of motion, opacity a function of division and so on, the outward qualities of bodies, in general, are functions of the motions and arrangements of their minuter parts. On the other hand, should such questions fail to get affirmative answers in an empirical investigation, it does not follow that the Aristotelian system stands proven, for it is only one among the logically possible alternatives. It was a lot easier to raise objections to the system of Aristotle than it was to devise an alternative which can be established beyond any scintilla of doubt.

Empirical investigations do not exhaust the kinds of question raised by a conceptual crisis. There are also questions of a more metaphysical kind. Empirical investigations are not immediately relevant to these, nor do empirical facts provide conclusive answers. An instance of the kind of question which might usefully be distinguished as metaphysical is whether matter is to be regarded as continuous or discontinuous. This is not an empirical question in any direct sense since a discontinuist could always push his claim back beyond the boundary of experimental discovery of continuity, and the continuist could, and in fact did, fill the interstices of granular matter with aethers. Nor is it purely philosophical, for it is not just a question of finding the most convenient meanings for the three expressions, 'matter', 'continuous' and 'discontinuous'. What is really at issue is logical adequacy; for instance a continuist could try to demonstrate inconsistencies in the concept 'discontinuous matter', that he believes his own alternative, 'continuous matter', to be free from. A good deal of the long discussion of this topic, initiated in antiquity by Zeno, has been aimed at defending one or other position against a charge of inconsistency, and counter-attacking with the same charge against one's opponents. Of course, much in these discussions depends on answers to philosophical questions, that is what meanings to assign to the key terms, but empirical facts cannot be entirely ignored in the matter. The pattern of argument is rather like this:

The internal consistency of some disputed position—say, discontinuity of matter—is argued for, usually by means of characteristic philosophical argument.

Assuming that consistency to be demonstrated, various empirical facts are adduced to demonstrate that the new position finds support in nature.

This philosophical service, the demonstration of freedom from logical contradiction, was performed for the corpuscularian philosophy by Pierre Gassendi. To demonstrate the possibility of atoms he demonstrated the possibility of the void, and to do this he had to separate from their traditional adherence the concepts of matter and space. Gassendi did not show that there was empty space, but he did make a pretty successful attempt at showing that 'empty space' is not a self-contradictory notion and so freed the concepts of space and matter from logical inter-dependence. A variety of philosophical techniques appear in Gassendi's arguments, ranging from Aristotelian logic-chopping to some fairly shrewd and modern-sounding linguistic analysis.

Gassendi heads this section of the *Syntagmata* 'de loco et tempore seu spatio et duratione' distinguishing place from space, and time from duration. In modern terms we would put the distinction he was after thus: that space is the class of places (whether occupied or not) and that duration is the class of times (whether anything happens in them or not). We need not consider the analysis of time and duration, which, though interesting in itself, does not contribute in any direct way to the foundations of the corpuscularian philosophy.

Traditionally, Gassendi points out, all *being* had been divided into the two categories 'substance' and 'accident'. What is neither substance nor accident is nothing. So we are tempted to argue that since space exists it must be either substance or accident. Even if we adopt the latter disjunct, since accidents must inhere in some substance, space as an accident would not be independent of matter. But, says Gassendi, it is quite clear that space is neither substance nor accident, and it is nevertheless real. Something must go: better the theory than the plain facts. Hence we must conclude that the traditional classification of being is wrong. In fact we need a threefold classification with a special category for space and time. Being is either substance (*quae per se sunt*), accident (*quae per aliud sunt*), or space and duration.

What shall we say about this third category of being? Space is certainly *per se*, it depends upon nothing else for its existence, but let us not be tempted for that reason to call it a substance. If we do fall into this temptation the very name we use will mislead us since the word 'substance' invariably suggests 'corporeal substance' which is just the mistaken view of space we are trying to escape from. The Aristotelian category connected with space is quantity, and space is continuous quantity. This is perfectly satisfactory provided we are aware of another temptation which lies in wait for us here, for we are tempted to ask 'Since space is quantity, quantity of what?' But this question is itself mistaken. 'The tyranny of matter', as Gassendi puts it, leads one to say that all quantity is *accidens corporeum*, and so to ask a question which presupposes that space is a quantity of something. Two moves can be made to free us from this tyranny.

(1) To the objection that a quantity of nothing is nothing, the apparent alternative to saying that space is a quantity of something, Gassendi replies that this nothing is the result, not of negation, but of abstraction. In received philosophy it is held to be legitimate to reach the concept of matter *per se*, by abstracting from any form all matter. If this move is legitimate then it must be equally permissible to abstract from any matter all form. In this way we reach the concept of form *per se*. Form, in this sense, is space. So if matter *per se* is a logically possible concept, so is the concept of the void. Admittedly space is imaginary, but it is not imaginary in the sense of being unreal or fictitious—rather in the sense of having to be constructed by analogy. We understand what a space devoid of matter is like by considering any corporeal body which would occupy it. We get an intuition of space only by imagining it to be corporeal, to be filled with matter.

(2) If the formal steps and the idea of construction by analogy do not secure conviction the argument can be reinforced by a little linguistic analysis. The word 'corporeal', as it is applied to accidents, means 'dependent upon body', as, for instance, 'length' and 'weight' are corporeal accidents—

properties dependent upon body. But space does not require this reference to body. It is, as Gassendi says, 'more than so much room'. It is not only the place of things, but the place for things. This allows us to say that there can be quantity without the actual existence of body. There is a quantity, space, where a body might be. There is not, however, a length which a body might have had, other than in the sense of a body which is imagined and can therefore be imagined to be of such and such a length. If we can have quantity in this hypothetical way then we can say that 'space is a quantitative reality independent of actual matter'. If we can say this without contradiction, then we have demonstrated the logical possibility of the void, and with it the logical possibility of corpuscles, material particles separated by empty space. Are corpuscles atoms? Are they the ultimate divisions of matter? Gassendi did speak in that way, but the more cautious of his contemporaries, and those who followed him in propounding this kind of conceptual system, on the whole preferred the expressions 'corpuscles' and 'particles', not committing themselves to any doctrine of the ultimate indivisibility of matter.

It is not important whether the arguments I have sketched above are cogent or whether there are flaws which one might want to attack. I am not defending corpuscularian empiricism. What does matter is that after Gassendi most philosophers and scientists (Descartes excepted), even though they may have opposed atomism, nevertheless thought the question to be one of empirical fact, and not to be settled by exposing the logical incoherence of the atomic doctrine.

With the corpuscularian philosophy established as not incoherent, that is not self-contradictory, the next step is to devise a way or ways of finding out, or proving hypotheses about, what properties the corpuscles actually have. In terms of the analysis in Part 1, Gassendi's arguments about space and matter show that the hypothesis that the basic individuals are corpuscles of matter is not unreasonable. It now remains to find out what are the basic properties of matter, and hence what are the essential properties of the basic individuals. Though Gassendi does not provide an answer to the question, he does outline a recipe for finding answers to questions of this kind. Nobody in the period of discovery that followed Gassendi's time paid much explicit attention to his recipe, but it is such a clear statement of what is required in trying to penetrate beyond the surface phenomena of nature that it is worth brief mention.

Gassendi makes use of three key concepts—*signa*, *occultae* and *indictiva*.

Occultae: these are objects, properties, processes or events in nature which cannot be observed, but which can be inferred from what can be observed. Gassendi believed that not only can we infer the existence of certain unobserved objects from observations of other phenomena, but we can also infer some, at least, of their properties. Corpuscles, their properties and the arrangements they take up in masses of matter are typical *occultae*. To find out something we do not know we always require what Gassendi calls a 'criterion' or 'instrument'. For instance, if we do not know the numerical measure of the length of an object we can find out what it is by using a foot-rule. The foot-rule is the instrument by which we pass from the known to the unknown. Such items are mechanical instruments, but there are also, by a most odd comparison, what Gassendi calls 'natural instruments'. In inferring to *occultae* we pass from the known to the unknown, so we will need an instrument. This is what he calls the natural instrument. Failing the microscope the only way we can get to the *occulta* in question is by making an inference from an observation by means of a conditional proposition which has the description of that observation either as its antecedent or as the negation of its consequent. We can be said to apply the instrument when we infer the consequent by *modus ponens* or the negation of the antecedent by *modus tollens*. The 'natural instrument' is the conditional proposition which carries us from the known to the unknown.

Signa: in employing the natural instrument we make use of

known facts, either for the affirmation of the antecedent of the conditional proposition, or for the denial of its consequent. Facts employed in this way are called by Gassendi *signa*. *Signa* are not confined to the inference of *occultae*. For instance, Gassendi gives the example 'If the sun shines it is day' and calls 'The sun shines' a *signum*. He defines *signum* quite generally. 'A fact which is known and used as the key to further knowledge is called "*signum, medium, seu argumentum*".'

Signa fall into a fourfold division: they are necessary or probable; they are *communefactiva* or *indictiva*. They are necessary when there can be only one *occulta* of which they are the sign, probable when there is more than one possible origin for them. For instance, if there is smoke there must be fire, since, for Gassendi, there is no other way of making smoke but by fire. But if there is death there only may be wounds, since there are more ways of dying than by the sword. Both necessary and probable *signa* are classifiable as either *communefactiva* or *indictiva*. *Communefactiva* are those *signa* which are the keys to objects which are possible objects of experience. For instance, when a man sees smoke he infers the presence of fire, and can proceed to verify this inference by actual experience. But much more important for science are *indictiva* since they are used as keys to *occultae*. In discussing the use of *indictiva* Gassendi recommends the following formula:

'If A is not then B is not, but B is, therefore A is.'

That is, *occultae* are reached as the necessary conditions for the occurrence or existence of those things or events which are treated as *signa*. They are not sufficient conditions. This has the advantage of allowing a step-by-step penetration of nature, since the attempt to find the sufficient conditions for phenomena in one intellectual act is too hazardous an undertaking. The example Gassendi gives of the working of his method is that of sweat proving the existence of pores in the skin:

'If there were not pores then men would not sweat, but men do sweat, therefore there are pores in the skin.'

Evidence for the properties and arrangements of corpuscles would carry conviction if set up in this form, but it was Boyle and not Gassendi who made the attempt.

Gassendi's method, and any other of a similar kind, depends upon the scientist who uses the natural instrument being in possession of genuine *signa*. Should these be in doubt nothing can be inferred of the *occultae* of which they are alleged to be the outward and visible consequence. Gassendi considers the objection that since a thing can appear or present itself differently to different people, no one of these presentations is the truth, and hence no one can be a reliable *signum*. He answers 'Suppose an object appears to me to be of a certain colour, I cannot say that this is "*ipsissima qualitas quae sit in objecto*", but I can say that this is *an* affection due to this object, or this is the relation that object realizes with *me*. The statement that the object is not red to you cannot make it cease to be red to me, while, on the other hand, the fact that it is not to you what it is to me proves that it is what it is, for if there were no objective reality the fact of difference could not be explained at all' (*Syntagmata*).

Illusions, differences in appearance, etc., are all *signa*, and can all lead to new knowledge about the world. This turning-of-the-tables is still perhaps the best answer to those philosophers who advocate phenomenalism on the basis of the argument from illusion.

Gassendi's method cannot lead to any decisions about the ultimate or most real constituents of the world. Empirical knowledge derived in his way is essentially hypothetical and always open to revision. After all, as Kepler pointed out in the *Apologia Tychonis*, it is wholly dependent upon the acceptability of the conditional proposition or natural instrument, one of whose elements is, in the interesting cases, always remote from perception.

More surety than this was needed and Bacon saw exactly how it was to be found. Expressed in the terminology I have adopted his merit was to have realized that some way must be devised for determining how far we have progressed towards a system of concepts which will be fundamental and hence universally employable. So ill has Bacon been served by most commentators, and so little has the irony of his choice of technical terms been appreciated that, though everyone knows of him, few can be said to be acquainted. To see his place in the pattern of the scientific revolution we must look closely into his philosophy of science.

BACON

In Bacon's *Novum Organon* there is to be found the first attempt at the invention of a method by which a general conceptual system could be established experimentally. What, in Part 1, I called the 'basic properties of basic individuals', Bacon calls 'the most real simple Natures'. One of the questions which the *New Instrument* is designed to answer is: which Simple Natures are the most real? Indeed the final aim of science for Bacon is to substantiate what I have called a general conceptual system in terms of which all natures other than those figuring in it can be analysed. This Bacon calls 'finding the Forms of Simple Natures'. To give an adequate account of his views it is necessary first to be quite clear as to what Bacon meant by his key concepts *Form* and *Nature*.

A Nature is a sensible quality: a colour, a sound, a sensation, a shape, etc. When we observe the natural world, what we observe are Natures, and when we make discoveries about the natural world what we discover are associations of Natures. To make anything of our discoveries we should present 'instances to the understanding' as Bacon puts it, by setting out what we know about the associations of Natures under three headings: (i) Agreement, instances in which two or more Natures occur together; (ii) Absence in Proximity, instances in which, where we would expect to find a particular association, one Nature is absent; (iii) Degrees, instances in which two or more Natures show mutual variation. A Baconian induction does *not* consist in rendering these associations general, as our present use of the word 'induction' would suggest. To understand what it was to make an induction, given such tabulated information it is necessary to understand the other central Baconian technical term, Form.

Bacon uses Form in several different senses,* but the central and crucial sense for his philosophy of science is that in which to find the Form of a given Nature is to give an analysis of that Nature in terms of the most real Natures. The statement of the Form of a Nature is the definition of that Nature in terms of those other Natures into which it can be analyzed. The object of the collection of instances of the associations of Natures in the tables of Agreement, Absence and Degrees, is to lead to the correct analysis and hence the most useful definition of the particular Nature which is the subject of the statements of instances of association. The first distinction among Natures which, according to Bacon, must be recognized, is that between Simple and Complex Natures. In each mode of sense a great many Natures are presented. For instance, to sight there are presented colours, shapes and textures. Of colours there are an infinite variety, including white, black and the chromatic colours. Newton's discovery that white is a mixture of the chromatic colours, would be expressed in Bacon's terminology as the discovery that White is a Complex Nature. The Form of White is, in the first instance, the set of chromatic colours needed to produce it. Furthermore, Newton's discovery of the unanalysability of monochromatic colours would have demonstrated for Bacon that these are Simple Natures. It is an induction, in Bacon's sense, to use the particular instances as set out in the tabular array, to discover the analysis of a Complex Nature in terms of some set of Simple Natures.

But this is only the beginning. To make a discovery of

---

* A somewhat similar use of 'Form' is to be found in Gassendi, who calls the 'inner constitution of a thing which enables it to act' its Form. *Syntagmata*, Physica, Sect. 1, Bk. IV.

fundamental importance, we need to find out what are the Forms of Simple Natures. Since we are presented with nothing but Natures in experience, the analysis of a given Simple Nature must be in terms of other Simple Natures, distinguished from the one to be analyzed by having, as Bacon puts it, more reality. We do not know *a priori* which simple Natures are the more real. We discover this by finding, in particular practical investigations, what Simple Natures turn up most frequently in the analysis of others. In the specimen tables which Bacon includes in his account of this process there are presented instances in which Heat agrees with other Natures, that is, is present with them; instances in which because of their similarity to cases of agreement, we would have expected to find the agreeing Nature present with Heat, it is in fact absent, and instances in which the degree of Heat varies with the degree of some other Nature. By scanning the tables in a certain way we make the induction, in Bacon's sense, that 'Heat is a motion, expansive, restrained and acting in its strife upon the smaller particles of bodies'. Since we can make the analysis of Heat presented in this definition, motion and configuration are Simple Natures more real than Heat. They can then be used to state the Form of Heat. So that in this instance the result of the tabular induction is not only an analysis of Heat, but also a statement of its Form.

The tables of agreement provide, in effect, a number of hypotheses concerning the Forms of the Natures under investigation. The table of absences in proximity enable us to rule out, one by one, those hypotheses which are not sufficiently general, since the absence of the associated Nature in situations where we would be likely to expect it gives us a strong counter-instance to the hypothesis. Thus we rule out as possible Forms those Natures and combinations of Natures which are associated only accidentally with the Nature under investigation, and hence do not contribute to its Form. Any hypothesis which survives does, provisionally at least, describe the Form of the given Nature.

Incidentally when this point has been reached we are not making an induction, in the modern sense of that word, that is, we are not making an inference from particular to general. What issues from a Baconian induction is a definition which is then put to use. While it continues to be effective no sensible doubts can be raised about its truth or its universal applicability, since for the purposes of its application it has become a necessary truth. If we do turn up an instance of absence in proximity to the definition or Form which we had been using satisfactorily up till then, the proper way of describing its effect upon that definition will not be to say that it shows it to be false, but that it shows the definition not to be universally applicable to phenomena of the given kind. That is to say, the definition which we had been using had not in fact reached the true Form of the given Nature.

A given Simple Nature is the more real as it turns up more often in the analysans side of the results of tabular inductions. This is how it is discovered which Simple Natures are the most real. But what has been discovered by this method? Can any more general philosophical conclusions be drawn from the repeated occurrence of a certain Simple Nature in the analyses of other, less real, Simple Natures? Bacon certainly thought that there were and here it is important to look at his distinction between physical and metaphysical enquiries, though he uses these words in a way of his own.

In physical sciences we find out particular causes for particular phenomena, but in metaphysics we find out the actual nature of kinds of phenomena. In metaphysics we find out, for instance, just what heat and light really are, while in physics we find out how to make things hot or luminous. As Bacon says,* 'Thus let the investigation of Forms, which are (in the eye of reason at least, and in their essential law) eternal and immutable, constitute *Metaphysics*; and let the investigation of the Efficient Cause, and of Matter, and of Latent Process, and the Latent Configuration (all of which have reference to

* Bacon, *Novum Organon*, Bk. II, p. 9.

the common and ordinary course of nature, not to her eternal and fundamental laws) constitute *Physics*'.

Using the method of tabular induction and the principle of exclusion particular causes can be eliminated and physics distinguished from metaphysics. For instance, Bacon gives the example 'Fire is the cause of induration, but respective to clay; fire is the cause of colliquation, but respective to wax. But fire is no constant cause either of induration or colliquation'. Fire cannot therefore be the Form of either induration or colliquation—that form must be something else. Causes and Forms must be distinguished, since Forms are, at bottom, identical with the Natures of which they are the Forms. 'When I say of Motion that it is the genus of which heat is a species, I would be understood to mean, not that heat generates motion or that motion generates heat (though both are true in certain cases), but that Heat itself, its essence and quiddity, is Motion and nothing else, limited however by specific difference . . .'*

Are Bacon's Forms identical with what later, by Boyle, Locke and Newton, came to be called primary properties? Bacon does not make this later distinction explicitly but there are plenty of hints in his methodological works that he was grasping after it. In the *Valerius Terminus* Bacon gives the following instructions for finding Forms in a way that clearly foreshadows the later distinction: 'that the nature discovered be more original than the nature supposed, and not more secondary or of the like degree . . . for the rule is that the disposition of any thing referring to the state of it, in itself or the parts, is more original than that which is relative or transitive towards another thing. So evenness is the disposition of the stone in itself, but smooth is to the hand and bright to the eye'.

In the *Novum Organon* (Bk. II. p. 13) he adds some further distinguishing marks and characteristics of the Forms which strengthen his claim to have anticipated much that has been put to the credit of Locke.

(i) The *forma rei* is *ipsissima res*, and the Nature and its Form

* Bacon, *Novum Organon*, Bk. II, p. 20.

differ only as *apparens et existens, aut exterius et interius, aut in ordine ad hominem et in ordine ad universum,* which is pretty much the way a complex of primary properties differs from the secondary quality we perceive. And in case one should think that appearance shuts us off altogether from reality he remarks 'Forms are not incognisable because they are supra-sensible'; though he does not repeat Gassendi's suggestions as to how they may be known.

(ii) Whenever Bacon actually produces any examples of Forms of Simple Natures, they are always given in terms of the arrangement and motions of the parts of bodies, as, for instance, in his famous definitions of heat as a 'motion, expansive and restrained' and of perceived whiteness as due to the mode of the arrangement of the parts of surfaces.

Bacon's account of scientific method and the goal of scientific enquiries brings out clearly that what is being sought is some kind of definition of matter, the substance of the corpuscles. Indeed the selection of certain properties to be the primary properties of matter is, or can be treated as, a search for a definition of matter. However, the kind of definition being sought is a little different from what we ordinarily regard as definitions. At least certain features of ordinary definitions are absent from the kind of definition being sought, while it has other features which our more usual definitions lack.

Definitions of matter are not to be classed with such word-definitions as 'congruence = df. exactly similar in all relevant respects' or with such species-definitions as 'hydra = df. multi-cellular, annulate, tubiform animal . . .' In each of these kinds of definition there is an element of arbitrariness and a degree of prescriptiveness that is absent in the definition of matter. The difference between the word/species type of definition and what I shall call quasi-definitions can be brought out in two contrasts.

(i) To say what matter is, is to propose an *analysis* of a concept already dimly or vaguely understood. This may involve either an excision of content from the concept as in the case of matter, or an addition to the content of the concept as in the case of the organic cell (where the growth of knowledge has led to a picture of a unit more complicated than had been imagined). In the word definition we start with, say, the concept of 'identical in all relevant respects' and find a word to express this complex concept economically; and thus we construct the meaning of congruent by definition.

(ii) There is always the logical possibility that the definition of matter could have been otherwise chosen, but this does not make the definition arbitrary, since what definition we do choose is forced upon us by certain general considerations, philosophical, empirical and pragmatic which are external to the definition itself; but that we call 'identical in all respects' 'congruent' instead of say, 'concomitant' is wholly arbitrary (etymological considerations apart), as is the decision to appropriate the name of a dangerous mythological monster for a harmless pond creature. Furthermore, that we should choose to treat as a singular concept 'identical in such and such respects' rather than 'identical in these and not identical in those' is arbitrary in the sense that we are not driven to this choice by anything other than our own decisions as to what is going to be of interest to us in geometry (compare the way we have chosen to retain the Pythagorean concept of a square number, but have dropped the concept of an oblong number, a concept just as important to Pythagoras).

To propose an analysis of a certain concept under the pressure of the general requirements of a general conceptual system towards which we have been driven by the accumulation of empirical, pragmatic and aesthetic considerations is to give what I have called a quasi-definition of a concept. The differences between quasi-definitions and word and species definitions which I have outlined above should not be allowed to obscure the fact that a quasi-definition has much in common with word and species definitions, with stipulative definitions generally. It does give a precision to a concept, though not arbitrarily; and it does have an air of analyticity, though its form does not follow from the meaning of its component words alone. It is not that someone like Newton says 'This is what I shall understand by *matter* from now on'. Rather it is like this: 'This is what the needs of science make it reasonable to suppose that matter actually *is*'.

BOYLE:
QUANTITATIVE CAUSES OF QUALITATIVE CHANGES

In his *Origins of Forms and Qualities* Robert Boyle set out the fundamental principles of the Corpuscularian Philosophy. As well as making perfectly clear what the scientific 'revolution' was establishing as a general conceptual system, he attempted to demonstrate the correctness of this system of concepts as the basis for the proper explanation of natural and artificial change. Boyle's principles are so clear in their statement of the general conceptual system that was in the process of establishing itself that they are worth quoting in full. In the *Origin of Forms and Qualities* (p. 35) he gives them in summary form as follows:

'(1) That the matter of all bodies is the same; namely a substance, extended and impenetrable.

(2) That all bodies thus agreeing in the same common matter, their distinction is to be taken from those accidents that do diversify it.

(3) That motion, not belonging to the essence of matter (which retains its whole nature when at rest) and not being originally producible by other accidents, as they are from it, may be looked upon as the first or chief mood or affection of matter.

(4) That motion, variously determined, doth naturally divide the matter it belongs to into actual fragments of parts; and this division obvious experience (and more eminently chemical operations) manifest to have been made into parts

exceedingly minute, and very often too minute to be singly perceivable by our senses.

(5) Whence it must necessarily follow, that each of these minute parts or *minima naturalis* (as well as every particular body made up by the coalition of any number of them) must have its determinate bigness or size, and its own shape. And these three, namely, bulk, figure, and either motion or rest (there being no mean between these two) are the three primary and most catholick moods of affections of the insensible parts of matter, considered each of them apart.

(6) That when divers of them are considered together, there will necessarily follow here below both a certain position of posture in reference to the horizon (as erected, inclining or level) of each of them, and a certain order or placing before or behind, or besides one another . . . and when many of these small parts are brought to convene into one body from their primary affections, and their dispositions or contrivance as to posture and order, there results that which by one comprehensive name we call the texture of the body . . . .

(7) That yet there being men in the world whose organs of sense are contrived in such differing ways, that one sensory organ is fitted to receive impressions from some, and another from other sorts of external objects or bodies without them . . . the perceptions of these impressions are by men called by several names, as heat, colour, odour; and are commonly imagined to proceed from certain distinct and peculiar qualities in the external objects which have some resemblance to the ideas their action upon the senses excites in the mind; though indeed all these sensible qualities, and the rest that are to be met with in the bodies without us, are but the effects or consequents of the above mentioned primary affections of matter, whose operations are diversified according to the nature of the sensories or other bodies they work upon.'

In the seventh principle we find the doctrine of primary and secondary qualities just as it is set out in Locke's *Essay*. However, the approach of Boyle to the proof or rather the advocacy of this doctrine is quite different from that which we will find in Locke. Boyle's method is to try to prove the general principle that qualitative changes in perceptual objects are generated by quantitative, motion and arrangement changes, in the corpuscles of the physical world. To do this he cites a number of experiments whose aim is to provide, in Gassendi's terms, necessary *signa* of the corpuscularian nature of matter. In contrast, Locke, as we shall see, attacks the problem in a kind of piecemeal way, seeking to show that each primary property has a perceptual counterpart in the perceived qualities of bodies, while ideas of secondary qualities do not have *other* properties as perceptual counterparts, but that the properties of matter which cause us to perceive secondary qualities, i.e. to have ideas of secondary qualities, are combinations of primary properties. For instance, the property of matter responsible for perceived colour is not a colour, but a texture.

To establish his conceptual system Boyle, in common with anyone of the period, felt obliged to attack and consequently to eliminate the doctrine of substantial forms, that structures and qualities are the result of the actualization of forms in prime matter, and particularly to eliminate the further implication of this doctrine that for each structure and for every quality there is an independent form. The most convincing examples of the operation of forms in nature are the crystals. Boyle cites the artificial production of crystals by chemical means as evidence that crystals are not the result of the actualization of a form in matter, but are structures built up by the aggregation of particles into certain characteristic arrangements. Summarizing his argument against substantial forms, he says,* 'If then these curious shapes, which are believed to be of the admirablest effects, and of the strongest proofs of the substantial forms, may be the results of texture; and if art can produce vitriol itself as well as nature, why may we not think that in ordinary phenomena, that have much less of wonder, recourse is wont to be had to substantial forms without any

* Boyle, *Origin of Forms and Qualities*, p. 59.

necessity? (matter and a convention of accidents being able to serve the turn without them); and why should we wilfully exclude those productions of the fire, wherein the chemist is but a servant of nature, from the number of natural bodies? And indeed since there is no certain diagnostic agreed on whereby to discriminate natural and factitious bodies, and to constitute the species of both; I see not why we may not draw arguments from the qualities and operations of several of those that are called factitious, to shew how much may be ascribed to, and performed by the mechanical characterization and stamp of matter.'

With the standard explanation of change out of the way and no longer constituting an obstacle to the acceptance of his own system, Boyle then proceeds to a positive account of the doctrine of Corpuscularianism and the explanation it gives of change. It is now in his dealings with qualities that Boyle's programme becomes evident. His aim, as has been pointed out, is the establishment of the general principle: *Quantitative Operations lead to Qualitative Change*. Four main examples are brought forward by Boyle to demonstrate his principle. It is mostly left up to us to draw the appropriate conclusions.

Example I: By distillation it can be shown that white of egg is similar in make up to other organic substances, so that in itself it has no special, unique features that make it suitable for the material out of which chickens are formed. It can be qualitatively changed by purely mechanical action as when it is turned from a 'tenacious' into a fluid body by the action of a whisk. What are we to say of the transformation of this substance into a chick? Boyle points out that new kinds of qualities appear in this substance as it is organized into a chick. There is the yellow colour of the skin and the red of blood, the hardness of bone and the springiness of tendons as well as more occult medicinal properties. Young magpies are a remedy for the falling sickness but white of magpie eggs is not. How has the egg white acquired these properties? The alternatives reduce to either the imposition of substantial forms or the re-arrangement of minute parts. According to Boyle, even if there is some principle in the cicatricula or speck which fashions the new matter, it must act in a physical fashion, for it has nothing else to act upon but the physical substance, white of egg. If this last proposition be adopted then we have an instance of profound qualitative change brought about by physical, and for Boyle that means quantitative, operations.

Example II: Growing plants in pure water provides another example of qualitative change. Boyle points out that growing plants in soil and adding only water does not prove anything, for the new matter may be hidden in the soil. But in nourishing vegetables with pure water 'all or the greatest part of that which would accrue to the vegetable thus nourished, would appear to have been materially but water with what exotic quality soever it may afterwards, when transmuted, be endowed'. The proof of this is that by distillation of a plant, so grown, a true oil can be obtained which will not mix with water.

Example III: According to Boyle the received opinion was that a plant or tree acquired the characteristics it has, not by transforming or rearranging material drawn from the earth, but by selection from among a stock of preformed substances already present in the earth. Against this opinion Boyle cites the example of grafting pear onto white-thorn stock. The fruit of the white-thorn is sour, of the pear sweet; but the pears grown on white-thorn stock are still sweet. Therefore the qualities of the fruit cannot be the result of a selection by the roots, since the white-thorn roots would have selected sour, not sweet.

Example IV: A number of chemical experiments are described by Boyle in which crystalline form, colour, texture and taste are changed by chemical processes, that is processes of solution, heating and so on. An interesting example is Boyle's reconstruction of the method of synthesis of Glauber's *sal mirabilis*; since from raw materials none of which taste like

Glauber's salt, the salt with its characteristic taste can be produced.

Boyle summarizes his investigations as follows: 'For, in the experiments we are speaking of, it cannot well be pretended, or at least not well proved, that any substantial forms are the causes of the effects I have recited; for in most of the (above-mentioned) cases, besides that in the bodies we employed, the seminal virtues, if they had any before, may be supposed to have been destroyed by the fire, they were such, as those I argue with would account to be factitious bodies artificially produced by chemical operations. And it is not more manifest that in the production of these effects there intervenes a local motion and change of texture by these operations, than it is evident and precarious that they are the effects of such things as the schools fancy substantial forms to be: since it is these new experiments, by the addition of some new particles of matter, or the recess or expulsion of some pre-existent ones or, which is the most frequent way, by the transposition of minute parts, yet without quite excluding the other two, that no more skilful chemist than I have been able to produce by art a not inconsiderable number of such changes of qualities, that more notable ones are not presented us by nature, when she is presumed to work by the help of substantial forms: I see not why it may not be thought probable that the same catholick and fertile principle, motion bulk, shape, and texture of the minute parts of matter, may, under the guidance of nature . . . suffice likewise to produce those other qualities of natural bodies, of which we have not given particular instances.'*

In explicit form the consequent reasoning must be taken to run in the following way:

(i) Quantitative operations lead to qualitative change: (this demonstrated by the experiments and examples just cited).

(ii) Quantitative operations are operations on the bulk, figure, motion and texture of the minuter parts of bodies.

(iii) Operations on such primary properties can lead only

* Boyle, *Origin of Forms and Qualities*, p. 112.

to new bulks, figures, motions and textures of bodies and their parts.

(iv) The consequent qualitative changes must then be *our* view of what are essentially bulks, figures, motions and textures of bodies.

The Necessary Properties of Matter

GALILEO AND THE METHOD OF CONTRASTS

Galileo begins his discussion of the distinction between primary and secondary qualities by proposing a quasi-definition of matter, in the following words: 'Whenever I conceive any material or corporeal substance, I immediately feel the need to think of it as bounded, and as having this or that shape; as being large or small in relation to other things, and in some specific place at any given time; as being in motion or at rest; as touching or not touching some other body; and as being one in number, or few, or many. From these conditions I cannot separate such a substance by any stretch of my imagination.'* He refers to this list of properties as 'necessary accompaniments' of the notion of material substance.

To this list of qualities which are alike perceived and also the properties of actual material substance, he contrasts another list of qualities which are perceived, but which, he claims, are not also the properties of actual material substance. He lists 'White or red, bitter or sweet, noisy or silent, and of a sweet or foul odour'. He claims that 'tastes, odours, colours, and so on are no more than mere names so far as the object in which we place them is concerned and . . . reside only in the consciousness'; that is, they are perceptual qualities only. The justification of this claim to separate perceptual qualities into two exclusive classes, one of which is constitutive of matter and

* Galileo, 'Il Saggiatore': In *The Discoveries and Opinions of Galileo:* Stillman Drake, p. 274.

the other of which is not, rests for Galileo upon what I shall call the *contrast argument*, an argument not studied so far as I know by philosophers. As I hope to show, the Galilean argument is susceptible of different exegeses only one of which will turn out to be satisfactory.

Let us first examine an exegesis of Galileo's argument which seems to be the most natural interpretation of it. It is to treat the argument as an attempt to establish a criterion for distinguishing primary from secondary qualities perceptually. Adopting this exegesis the argument can be put in the following form:

(1) Consider a certain mechanical interaction, say contact between two bodies. So far as the mechanical interaction is concerned it is the same wherever it occurs. In particular it is identical in contact between living and non-living bodies, living and living bodies, non-living and non-living bodies.

(2) However, there is a contrast between its effect on non-living bodies and on *certain parts* of living bodies. On some parts of living bodies the effect is the same as on non-living bodies, that is, purely mechanical. On other parts the effect is quite different; it is then mechanical action plus the sensation felt by the living body.

(3) It follows from this that the difference or contrast in the effect must be a function of the living body, not in so far as it is a body but in so far as it is living. The difference in the effect is a difference in the consciousness of the mechanical effect by the living body. Therefore the difference or contrast must be a function, specifically, of the consciousness of the living body, and hence a quality which exhibits the contrast effect must be at least in part a quality only of percepts and not a property of things as well, since so far as things are concerned there is no contrast, the motions or mechanical interactions being identical. Galileo's example is that of the effect of touching bodies and the production of a tickling effect in the living body, but only on certain parts. He suggests but does not construct, arguments to demonstrate a similar distinction between mech-

anical qualities that are also the properties of matter and 'tastes, odours, colours, and many more'.

Before considering the force of this exegesis of the argument in more detail it is worth noticing that it is not an argument which has engaged the attention of philosophers. This may in part be due to the fact that Locke does not repeat it and also in part due to the general neglect by philosophers of the philosophical arguments that were used by scientists in the empiricist tradition. Not only have the arguments of Galileo been neglected but there are also unique forms of argument used by Newton which have neither been repeated nor criticized in the works of professional philosophers since. Even Berkeley does not attack Newton's philosophy but Locke's. This can hardly be because the arguments of Galileo and Newton lack merit, but may be due to the fact, remarkable if true, that empiricism was advocated, condemned and disputed by generations of philosophers who seldom, so far as one can judge, made themselves thoroughly acquainted with the works of empirical science for which that philosophy was the ultimate justification.

Galileo then distinguishes perceptual qualities which are also properties of matter, and perceptual qualities which are not also properties of matter but which would, as he says, not be held to exist if there were no organs of sense. (This remark it must be noticed would be remarkably unacute if we adopt this exegesis, since pursuing this point would drive us to conclude that the primary qualities could not really exist either since it would be essential to the argument, understood in this way, that it is a distinction in classes of perceptual qualities that we are making.) The conclusion Galileo wants to reach is that as far as the object is concerned, the latter class of qualities are mere names. The argument, as it has been presented according to this exegesis, clearly will not do, for, though we could make it cogent for such qualities as tickling and heat, both of which show variability over the body when the physical accompaniment of the sensation is identical, the contrast or

G

variability criterion which we are supposing Galileo to be proposing could not distinguish colour from shape, for colour and shape when perceived by the eye are perceived *together*. We simply do not get the kind of variability, i.e. spatial variability with constancy of mechanical accompaniment, that Galileo's contrast argument needs.

To generalize the argument to apply to colours it would be necessary to consider not only variability over the expanse of the body, that is spatial variability, but variability with time. A square cloth which also looks yellow at midday, is still square but looks grey at dusk. Even this fact will not do the job for Galileo since the essence of the argument is to contrast the variability of the perception with the constancy of the mechanical interaction which accompanies it. At dusk the absence of the sun would have to count against the constancy of the mechanical accompaniment because the difference in the percept could be attributed to a difference in the objective set-up which includes both the object perceived and the sun. The kind of argument which might do for colours would be to consider the effects of drugs or illnesses like jaundice but in which there is general agreement among other observers that no significant change has occurred in the rest of nature.

Even if this extension is admitted a converse objection can still be raised against this exegesis. It is clear that the argument will not provide a criterion, without some modification, for all the qualities which Galileo wishes to have as secondary. But equally one might object that variability of sensation with constancy of mechanical accompaniment is inadequate as a criterion since even the qualities which Galileo wishes to have as primary are also variable. Shape, for instance, shows a well-known variability with the position of the observer as does motion with the relative motion of the observer. Cases could no doubt be constructed to demonstrate variability with all Galileo's primary qualities. It looks, therefore, as if the simple contrast argument which distinguishes perceptual qualities as 'mere names' if they vary to the perceiver against a constant mechanical accompaniment will not provide an adequate criterion. It fails either:

(a) Because it is extremely difficult to demonstrate the appropriate contrast with some of the properties Galileo wishes to locate only in the percept; or

(b) Because many of the perceptual qualities which Galileo wishes to recognize as also properties of matter are also variable.

The latter objection is not so strong for (a), for the criterion proposed is not simple variability but variability in contrast to constancy of mechanical accompaniment. Furthermore, there would be a certain naivety in accepting too readily the counter argument about the variability of shape. It is to be emphasized that we are contrasting perceptual qualities of objects and not simple sensations, or bits of our visual fields. It is a demonstrable psychological fact, often overlooked by philosophers who worry about the possibility of confusion between the elliptical shape of a penny seen at an angle and ellipses seen straight on, that we can and do quite easily *perceive* the difference just as we perceive the difference between square tables seen at an angle and rhomboids or diamonds seen straight on. Similarly as Gombrich* has pointed out with copious illustration, we *perceive* the same colours and tonal contrasts in dirty and cleaned pictures, though our *sensations* of colour and tone are in each case quite different.

To treat Galileo's argument as a method for establishing a criterion for distinguishing two kinds of perceptual qualities is not successful. But it can be looked at in another way. It could be an illustration of what can be achieved by adopting a certain general conceptual system expressed in the list of properties which Galileo finds to be inseparable from and necessary accompaniments of the notion of matter. Having adopted the general conceptual system we can then see that certain other qualities are variable in contrast to the quality/ property of the general conceptual system. Adopting this view

* Gombrich, *Art and Illusion*.

of the matter the argument must then be seen, not as a method of establishing criterion for distinguishing two kinds of qualities, but rather as a way of backing up a certain selection of qualities to constitute the quasi-definition of matter.

We can look upon Galileo as proposing a certain quasi-definition of matter and then as using the contrast argument to show that the quasi-definition works, that we do not need any more properties for matter than those proposed. What the contrast argument can be thought to demonstrate is that all other perceptual qualities are not also properties of matter because it is reasonable to suppose that they are the effects of certain combinations of the quasi-definitional set of properties on our consciousness. From this point of view the tickling argument is a sound one, for the tickling sensation can only be a product of the mechanical movement of the hand or feather and the sensitivity of the subject. Galileo cites similar reasons for deciding that tastes, smells and sounds are wholly perceptual. The reason why these arguments are successful is that the constancy of mechanical accompaniment of the sensation is demonstrable. We can observe the grains of pepper, or attar of roses, and easily demonstrate that we cannot smell or taste with our hands or feet, but only when these substances come into contact with the appropriate sense organs do we experience the sensation.

With sensations of heat we can demonstrate the variability easily enough, but we cannot very well identify the constant mechanical accompaniment, except by a step of theoretical invention. This is, in fact, Galileo's method. Having established both the variability of sensation in the percept, and the constancy of mechanical accompaniment in the objects for touch, taste, smell and sound, he then takes the theoretical step of adducing the same sort of constant mechanical accompaniment for sensations of heat in the perception of hot bodies. This way of interpreting the argument is supported by Galileo's method of introducing the subject of heat. He says, 'having shown that many sensations which are supposed to be qualities residing in external objects have no real existence save in us, and outside ourselves are mere names, I now say that I believe heat to be of this character'. He goes to construct in theory a machinery for the production of sensations of heat, as follows: 'Those materials which produce heat in us and make us feel warmth, which are known by the general name of "fire", would then be a multitude of minute particles having certain shapes and moving with certain velocities. Meeting with our bodies they penetrate by means of their extreme subtilty, and their touch as felt by us when they pass through our substance is the sensation we call "heat".' So while admitting that we cannot demonstrate experimentally this theory of heat, Galileo is trying to put forward, on the analogy of the demonstrable theory of the nature of tickling, a corresponding theory of heat. It is a theory constructed from the same general conceptual system as the theory of touch, taste, smell and sound. The success or reasonableness of the new theory of heat is a backing up for the adoption of the general conceptual system, and in so far as every phenomenon can be ascribed to the operation of mechanisms constructed out of materials admitted in the general conceptual system, that g.c.s. is 'proved'.

In summary then we find that though Galileo seems at first sight to be proposing a criterion for distinguishing perceptual qualities which are also properties of matter, from perceptual qualities or sensations which are not, this exegesis will not do. However, if we understand the argument as a demonstration of the efficacy of the general conceptual system for explaining some perceptual qualities not mentioned in the general conceptual system lists, and then as a hypothetical generalization of the demonstrated result to other qualities not mentioned in the lists of the general conceptual system, the argument may lose something in strict validity but gains much in power of conviction and plausibility. 'Proving' the general conceptual system then becomes the fulfilment of a programme aimed at showing that it is reasonable to understand all perceptual qualities not mentioned in the general conceptual system as the effect on our sensibility of properties which are in the general conceptual system, where this effect is not a quality or group of qualities identical with the property or group of properties which produce it. This argument can be produced in general form independent of any particular conceptual system.

Let the list of qualities given in perception, or detectable by instruments, be

$$p_1...p_t.$$

Then let it be demonstrated that some subset of these properties,

$$p_k...p_m$$

are the product of human sensibility with some other subset of qualities, say

$$p_1...p_{k-1}.$$

We then adopt the hypothesis, H, that the properties of matter are the qualities $p_1...p_{k-1}$. Now, in so far as we can convincingly construct with $p_1...p_{k-1}$ accounts of how we come to experience $p_{m+1}...p_n$, we shall be justified in regarding H as better established the nearer n approaches t. There are endless examples of scientists of some particular era struggling to 'make n approach t'. For instance, the attempts in the nineteenth century to give mechanical accounts of electro-magnetism, the attempts in the twentieth century to give field accounts of all basic interactions, the attempts in fourth century B.C. to give developmental accounts of purely mechanical phenomena and so on.

In brief, one might say that the contrast argument in the end leads, not to a distinction among qualities but rather to a distinction among sciences. Since those qualities which are primary and which mirror or are the model for the real properties of matter are the subject matter of a science of bodies in so far as they are independent of human consciousness. This independence is of course not independence in general, but the particular kind of independence which the contrast argument demonstrates. (One might say that some modern phenomenological epistemology exists because philosophers have confused general and particular independence.)

JOHN LOCKE:
THE THEORY OF PERCEPTUAL COUNTERPARTS

If we accept Galileo's argument we are not thereby freed from all problems. One remains. It is the problem which raises itself as soon as we try to fill in, in any detail, some relation that might be supposed to hold between those qualities which are selected by some means as primary, and the properties of matter. Is there an identity between the properties of matter and the qualities so selected, or is there a 1-1 correspondence between these qualities and the properties of matter, or is perhaps the whole business of seeking for this relation useless? Neither Galileo nor Boyle really tackle this question, and without a solution or dissolution of it the methods of those two natural philosophers can hardly be said to demonstrate what the true or real properties of matter are if matter is treated as the cause of our perceiving. This question was tackled by John Locke, the physician, philosopher and friend of Newton and Boyle.

Locke set himself the task of developing in a coherent, systematic and rational way what he took to be the fundamental tenets of the corpuscularian philosophy and of thus establishing the Newtonian conceptual system upon a foundation of reasoned argument. That the Newtonian science, the laws of matter in motion which interacted by impact, was substantially correct Locke evidently took for granted. The programme of the *Essay* is an attempt to show why this system is correct and how it can be that the Galilean selection of qualities does give us the decisive lead to the properties of matter. To establish his conclusion Locke needed to demonstrate the following propositions, though he did not attempt their demonstration in quite this order.

(1) All versions of the Aristotelian general conceptual system are useless for science, since both the fundamental concepts of that system are defective—'substance' because it cannot be detected empirically and is never, by the very nature it is alleged to have, the object of perception. Boyle, though he uses the term, avoids this difficulty in setting out the corpuscularian principles because he turns from 'substance' to the 'something extended and impenetrable' as constitutive of body; 'forms' because the forms of all perceptual or instrumental qualities are of equal status, while successful science, as demonstrated by Boyle and Newton, works by recognizing only a limited subset of qualities of bodies as corresponding in some essential way to the constitution of body and hence as being guides to the real properties of matter.

(2) Locke's own solution to the problem of the relation between perception and reality is to argue that a distinction can be made in the set of perceptual qualities and reactions of instruments dividing those that are in 1-1 co-ordination (resemblance) with properties of matter from those which are perceptual or instrumental only.

A point about the way Locke's written works should be treated needs to be made before proceeding to an exposition of his philosophical opinions. The *Essay* was built up from notes and fragmentary papers composed and collected over a long period. This method of composition has the effect that the doctrine which Locke needs to present is not always clearly or steadily presented. Sometimes vital distinctions are not adhered to by Locke as firmly as consistency would demand. For the purpose of this study I do not propose to confront Locke with his own slips and inconsistencies but rather to extract from the *Essay* as consistent and coherent a version of the empirical philosophy as I can find in it, without, if possible, putting more into the system than can be found in the *Essay*.

Ideas, Qualities and Properties. The distinction between qualities of percepts and properties of objects which I have been using in explaining Galileo's remarks on his general conceptual system is essential to making those remarks clear but was not made explicitly by Galileo. Locke makes the distinction entirely explicit and quite clear, though his use of the distinction is not always quite coherent. He distinguishes accidents 'as they are ideas or perceptions in our minds, and as they are modifications of matter in the bodies that cause such perceptions in us'.* Locke nowhere makes any claim to know how the modifications of matter can cause us to have sensations and to construct percepts out of them, but he does assume that it is quite obvious that they do. This assumption is a matter of some importance. It can be tackled in a variety of ways. We could, for instance, examine the effects of denying it upon our language and what we can be said to know. Another way of dealing with it is to try to demonstrate that the problematic air of the question 'How can modifications of matter cause sensations in us?' is removed when we come to see that sensations are also modifications of matter. They seem to be something different only if we fail to notice that this difference is the difference between what one might call 'inside' and 'outside' views of modifications to those complex giant molecules we call human beings. Let us then leave aside this problem in the meantime, since without the assumption of percepts being due to the interaction of organism and environment we can hardly begin.

Locke distinguishes percepts and sensations from modifications of matter in the following way: 'Whatsoever the mind perceives in itself, or is the immediate object of perception, thought or understanding, that I call an *idea*, and the power to produce any idea in our minds, I call a *quality* of the subject wherein that power lies'.† Among perceptions we can find no further distinction than the differences exhibited by the ideas themselves. Locke says, 'the ideas of heat and cold, light and

* Locke, *Essay*, Bk. II, Ch. VIII, § 7.   † *Ibid.* § 8.

darkness, white and black, motion and rest, are equally clear
and positive ideas in the mind. . . . These the understanding,
in its view of them, considers all as distinct positive ideas without taking notice of the causes that produce them.'* The study
of the ideas as we experience them, and the explanation of why
we perceive what we do are two totally different kinds of enquiry. As Locke says 'it being one thing to perceive [*sic*] and
know the idea of white or black, and quite another to examine
what kind of particles they must be, and now ranged in the
superficies to make any object appear white or black'. Locke
does not make this distinction as a dogmatic assertion but
adduces a reason for this view. The 'ideas' he cites as examples
are paired; heat/cold, light/dark, white/black. The first member of each pair was held to be caused by a real, positive property in things, while the second member of each pair was
perceived, not when the object had some other positive property but when the positive property causing the person to
perceive the first member of a pair, is absent. But when we
perceive the first member we do not perceive nothing at all,
we perceive a distinct, positive quality in the object. It follows
then that if the causal hypothesis about perception, that what
perceptions we have are at least partly the effects upon us of
things other than ourselves, is acceptable, there are positive
qualities for which there are no corresponding properties in
things. It further follows that many perceptual qualities must
be regarded as different, unless it be proved otherwise, from
the real properties of things.

The corpuscles, however, must have some properties, and
we might regard Locke as proposing an hypothesis as to which
qualities these properties correspond, deriving his list in the
first instance from the properties which were required by
Newton in his science of matter. The rest of the discussion can
be regarded as the attempt to show that the joint hypothesis
that these are *the* properties of matter and that there are corresponding qualities, is correct. That is, I think it would be a

* Locke, *Essay*, Bk. II, Ch. VIII, § 2.

mistake to read Locke as proposing an analysis of experience which anybody, at any time, might have made if only they had attended carefully enough to the facts, but rather as proposing an hypothetical g.c.s. suggested by the success of the sciences of the day, and then looking for a justification of this general conceptual system in ordinary experience as well.

This complex hypothesis can be put in the following form:

(1) Among our ideas (qualities of perceptions) we distinguish ideas of primary qualities and ideas of secondary qualities.

(2) The 'ideas' of primary qualities are perceived extension, perceived solidity, perceived figure and perceived motion; the 'ideas' of secondary qualities are perceived colour, perceived warmth, perceived taste and so on. It is a mistake to father on Locke the doctrine that colours, warmths, tastes etc. are secondary qualities, it must be insisted that like all that is perceived they are 'ideas' and thus are ideas of secondary qualities.

(3) The primary qualities (properties) of bodies are extension, solidity, figure and motion. The secondary qualities of bodies are various combinations, collocations and resultants of primary qualities (properties), e.g. 'the arrangement of particles in the superficies of bodies'.

(4) To use my own terminology now, the qualities extension, solidity, figure and motion are those qualities of percepts which are intimately related to the properties of objects since there is supposed to be 1-1 correspondence of 'resemblance' between such properties and qualities; but, e.g., 'the arrangements and motions of particles' are properties of bodies, but are perceived as those qualities which we distinguish as colours, warmths, textures and so on.

To establish these hypotheses as the basis of science Locke proceeds through the following steps:

(i) The general notion of 'substance', though present in the corpuscularian conceptual system as outlined by Boyle, is to be eliminated *from science* in favour of what Locke rather disingenuously calls 'the common notion of substance'. This common notion is identical with Galileo's quasi-definition of matter.

(ii) The recognition of primary properties is shown to be quasi-mechanical in that we come to perceive the corresponding qualities via some Newtonian interaction. If this is so then we have good reason to suppose, Locke suggests, that any division of matter which can act mechanically must have the primary or mechanical properties and hence, if we could perceive it, would be perceived to have the primary qualities.

(iii) If mechanical interactions are real, and hence if mechanical properties are real, then secondary qualities do not resemble the properties which produce them.

(iv) When talking of *things* we are now in a position to distinguish the nominal from the real essences of things; and following from the distinction we can construct a programme of science that has some chance of being fulfilled.

Substance and body. If we put the question, 'What is the world made of?' to an Aristotelian, he must reply, Substance. And if we ask him, 'What is substance?' he must reply, 'That in which all accidents inhere'. Substance is the stuff which itself has none of the qualities we perceive in things, but can acquire any and all of such qualities. Clearly 'substance' in this sense cannot be perceived. As Locke put it 'so that of substance we have no idea of what it is, but only a confused, obscure one of what it does'. This being the case 'substance', in the philosopher's sense, can play no useful role in science. But there is something, akin to substance, which does play a role in Newtonian science, and Locke calls it *body*. We get at it through what Locke calls 'the common notion of substance'. This consists of complex ideas of three kinds.

(1) 'The ideas of primary qualities of things which are discovered by our senses, and are in them even when we perceive them not: such are the bulk, figure, number, situation and motion of the parts of bodies'.* At this point Locke does not

* Locke, *Essay*, Bk. II, Ch. XXIII.

need to argue for the implicit assumption of resemblance between our ideas of primary qualities (qualities) and primary qualities (properties), for here he is, according to my interpretation, expounding his hypothetical conceptual system as it affects substance.

(2) 'The sensible secondary qualities which, depending on these [the primary qualities] are nothing but the powers those substances have to produce several ideas in us by our senses, which ideas are not in the things themselves, otherwise than as any thing is in its cause'. 'The sensible secondary qualities' no doubt form one of the 'common notions of substance' but the exegesis of what such secondary qualities are is far from common. At this point in the argument we must treat it as an hypothesis yet to be demonstrated.

(3) 'The aptness we consider in any substance to give or receive such alterations of primary qualities as that the substance so altered should produce in us different ideas from what it did before.' Again we are presented with an hypothesis (Boyle's Principle), which, even given the cogency of the primary and secondary quality distinction is yet to be confirmed. That of which the three hypotheses above are true is 'body', and it provides Locke's version of the Newtonian answer to the question, 'What is the world made of?' It has this advantage over substance, that it is not devoid of recognizable properties. Indeed it is such that given a proper understanding of what perceptual qualities are, namely for Locke, ideas in the mind caused by the interaction of things with sensitive organisms, it has both primary and secondary properties. In sum, 'the primary ideas we have *peculiar to body* . . . are *the cohesion of solid* and consequently separable *parts, and a power of communicating motion by impulse* . . . figure is but the consequence of finite extensions'.*

The Recognition of Primary Properties. One way of establishing the reality of any property corresponding to a particular sensation, in itself part of the analysis of a perceptual quality, would

* *Essay*, Bk. II, Ch. XXIII, 17.

be to try to show that the sensation is related to the property in such a way that the property must have some kind of identity, though not necessarily qualitative, with the corresponding sensation. The empiricist procedure here is something like the way the Action and Reaction law is used to deploy the formal concepts of mechanics. If a body is suddenly seen to acquire momentum, and we know the Law of the Conservation of Momentum, we can be sure that some other body, *having momentum*, has collided with it. Locke employs the model like this: if we feel a resistance with our fingers then there must be an equal and opposite resistance in that with which our fingers are in contact. Locke argues* as follows: 'The idea of solidity we receive by our touch: and it arises from the resistance we find in body to the entrance of any other body into the place it possesses, till it has left it . . . that which thus hinders the approach of two bodies when they are moving one towards another I call *solidity*. . . .'

An assumption of the greatest importance is concealed in this argument. It can be put in two ways:

'Our senses can be treated as instruments'
or
'Parts of our bodies are not privileged objects.'

If we observe that, say, the action and reaction equality holds between two perceived objects independent of our bodies, then, adopting the second form of the above principle, it will also hold for contact between any perceived part of our own bodies and an object. In particular if we observe that any two bodies offer resistance to mutual penetration, the same resistance is being offered by a body and our hand when they are in contact but resist mutual penetration. Therefore the resistance we perceive is the *perceptual counterpart* of a real resistance on the part of the body that we touch or push against. The perception Locke calls 'resistance', and the primary property of which it is the perceptual counterpart he calls 'sol-

* In Bk. II, Ch. IV, of the *Essay*.

idity'. Primary qualities are the perceptual counterparts of primary properties, but secondary qualities are not the perceptual counterparts of further, real properties, they are the perceptual counterparts of certain combinations of primary properties.

This is the sense in which the perceived primary qualities 'resemble' the real primary properties of bodies. It is not, what many critics and commentators on Locke seem to think, that the property and the quality are given qualitatively identical but that the quality is the perceptual counterpart of the property. We can perceive the qualities which are the perceptual counterparts of primary properties, and we discover that the objective counterpart of the sensation, quality or idea of resistance is mechanical impenetrability, that is physical resistance. Both real property and perceptual counterpart are resistances. But with secondary qualities this is not so. The objective counterpart to chromatic qualities is not colour, but an arrangement of atoms, or an activity within them.

Body being defined by primary properties in this way, derived as the objective counterpart of our idea of body in Locke's terminology, or of the perceived bodies in mine, can we set any lower limit to the size of bodies? Bodies we perceive, that is in Locke's terminology, ideas of bodies, can be divided and subdivided up to a point, beyond which we cannot distinguish the division, but as far as we can go the subdivisions of body exhibit all the defining characteristics of body as we perceive it before we divide. Locke argues for the universality of the primary properties in subdivision of the objective counterpart, however far it is divided. To justify this Locke falls back upon the quasi-definition of body. This suggests that he is not really setting about proving that this is what body must be, but demonstrating the reasonableness, not the inescapability, of a pervasive, methodological hypothesis, much in the manner of Galileo. Locke says that though our senses take no notice of solidity, 'but in masses of matter, of a bulk sufficient to cause a sensation in us; yet the mind having once got the idea from such grosser sensible bodies, traces it further, and considers it, as well as figure, in the minutest particles of matter that can exist, and finds it inseparably inherent in body, wherever or however modified.'*

Nominal and Real Essence: The Programme for Science. What then, by employing this general conceptual system, can we be said to know, and what can we discover? What we know now by means of our senses Locke calls the *nominal essences* of things. We can know what the primary properties of bodies are since they are the objective counterparts of primary qualities, and we can know what secondary qualities are found with what primary qualities. To know these collocations of qualities is to know the nominal essences of things: 'nominal' since it is these collocations of qualities that we use in the first instance to recognize things, and hence to give sense to the names of things. But to know the *real essences* of things we should need to know not only what primary properties a thing has, but also what are its secondary properties, that is what combinations, collocations or resultants of primary properties are the real objective counterparts of secondary qualities. Ideally, the programme for science is quite simple—get to know the *real* essences—and this is an identical demand with Bacon's, that we get to know the Forms of Simple Natures. The problem is to determine how far this ideal can be achieved.

(1) *What we can do*:

(*a*) Failing a means of discovering the minuter constitution of bodies, and hence while we are confined to unaided observation, we can develop only what I have called in Part 1 reticular theories. As Locke puts it, 'The advances that are made in this part of knowledge, depending on our sagacity in finding intermediate ideas that may show the relations and habitudes of ideas, whose coexistence is not considered'.

The contemporary, eighteenth-century situation was summed up by Locke as follows: 'By the colour, figure, taste and smell, and other sensible qualities we have as clear and distinct

* Locke, *Essay*, Bk. II, Ch. IV.

ideas of sage and hemlock, as we have of a circle and a triangle: but having no ideas of the particular primary qualities of the minute parts of either of these plants, and of other bodies which we could apply them to, we cannot tell what effects they will produce; nor when we see those effects can we so much as guess, much less know, their manner of production'.*

(b) But if the connection is made via 'the mechanical affections of the particles' of bodies we should be able to say how a great many things behave. As Locke says, 'Did we know the mechanical affections of the particles . . . the dissolving of silver in *aqua fortis*, and gold in *aqua regis*, and not vice versa, would be then perhaps no more difficult to know, than it is to a smith to understand why the turning of one key will open a lock, and not the turning of another. But whilst we are destitute of senses acute enough to discover the minute particles of bodies, and to give us ideas of their mechanical affections, we must be content to be ignorant of their properties and ways of operation.'

(2) *What we cannot do*: There is no limit to what we can achieve by investigating the primary properties of bodies, for to each there is a perceivable perceptual counterpart, the primary quality. But,

(a) However many correlations we can discover between certain arrangements and activities of primary properties and the idea of secondary properties or secondary qualities we cannot get any nearer to knowing how these sort of causes produce these sort of effects in us. There is a suggestion here that some kind of categorical disparity separates the collocations of primary properties that are the objective counterparts of secondary qualities and the secondary qualities in perception.

(b) It is not possible to get 'certainty and demonstration', that is 'general, instructive, unquestionable truth' or, as Locke rather paradoxically puts it, 'there is no science of bodies'. By this Locke means no more than what is usually meant by

H   * Locke, *Essay*, Bk. IV, Ch. III, § 26.

saying that every law of nature is stated under the limitation of induction, that is that we can have no *general* guarantee that the course of nature will not change radically at any time.

The Newtonian Synthesis

TRUE MECHANICAL MATTER

Newton's achievement in science is often described as a synthesis of all that was known before him. This is also true of his philosophy of science. It is not clear whether Newton was much acquainted with philosophy but in his own brief but important philosophical remarks he succeeded in picking out of the wavering and uncertainly controlled insights of his intellectual predecessors just the most significant of the doctrines towards which they had, with so much labour, been struggling. In establishing the new general conceptual system, three steps were required:

(1) The adoption of the corpuscularian philosophy, advocated as we have seen by Gassendi and Boyle.

(2) The selection from among all possible perceptual qualities of a set of properties both constitutive of matter, and extendable to divisions of that matter no matter how far continued. The argument towards the substantiation of the hypothesis that these are bulk, figure, motion and texture we have seen carried on by Bacon, Boyle, Galileo and Locke.

(3) The consequent choice of impact and, with great reluctance, gravitation, as the fundamental interactions between corpuscles constituted by the objective counterparts of the chosen qualities or primary properties of matter.

Bacon's criterion for the reality of a Nature or the primacy of a quality was that it should be found in the survivors of all tabular inductions. Galileo's criterion for the primacy of a quality was that it should be part of the quasi-definition of matter and that, for him, meant that it should not be variable in a way which depended on its relation with human sensibility. Locke we have seen making a distinction between qualities which have distinct and positive objective counterparts and qualities which do not. Summing up these distinctions we could say that if a quality is found as the survivor of all inductions, if it is definitionally constitutive of the notion of matter its primacy is recognizable by the *universality* of its appearance in the objects of that part of experience of which mechanics is the science. If, further, it is not subject-dependent, its primacy is further recognizable by its *invariance*. Newton, with his usual clarity of insight, choses just these criteria—universality and invariance—for making a selection of the significant or primary properties of matter, but uses them in his own special way. There are differences of the greatest importance between him and those who had made attempts to use these criteria before him.

(*a*) The invariance which Newton recognized was not that of subject invariance. It is a much more general invariance. He says, 'The qualities of bodies, which admit neither intensification nor remission of degree, and which are to be found within the reach of our experiments, are to be deemed the universal qualities of all bodies whatsoever.' Two points should be noticed about Newtonian invariance.

(i) Primary qualities are to be limited to some selection of those qualities 'within the reach of our experiments'. This insistence leads to the possibility that has been realized since Newton's time—that the set of qualities from which the primary properties of matter are to be drawn may not be co-extensive with the set of perceptual qualities. There may be certain perceptual qualities which cannot be made the subject of experiment and hence are ruled out as candidates for primacy, while there may be other properties of bodies, e.g. the magnetic, which are not perceptual qualities at all, but show up only in the behaviour of instruments. It is partly because of the importance of this development that I have chosen to introduce the quality/property terminology, even though it cuts across philosophic usage and runs the risk of breeding confusion.

(ii) 'Qualities which admit neither intensification nor remission of degree' might mean 'qualities which are subject-invariant', that is qualities independent of their physical environment; and which change neither with time nor distance. It is quite clear that Newton, in contrast to Locke and Galileo, means qualities invariant in the latter sense. For instance, I would be inclined to think that Newton would have regarded the colour of homogeneal light as being a quality which admitted neither intensification nor remission of degree, though since not all bodies are the same colour, hence not the colour itself but only the power to be coloured has some primacy.

(b) Newton, more than any other of the philosophers and scientists we have considered, was clear about the quasi-definitional character of the ascription of the primary qualities to objects, of the ascription of the primary properties to matter. For him universality is not an independent criterion, but dependent upon the appropriate invariance of a quality and its universality in perception. Invariant qualities are, he says, 'to be *deemed* the universal qualities of all bodies whatsoever' (my italics). How do we know that a certain quality is a primary property? It must be invariant in the Newtonian sense, and it must be 'universal in perception: . . . because we perceive it [extension] in all bodies that are sensible, therefore we ascribe it universally to all others also'.

(c) To complete his account Newton brings in another principle, deeply embedded in the whole of science. The principle is that large-scale phenomena are the resultants of groups of smaller-scale phenomena. Newton says, 'That abundance of bodies are hard, we learn by experience; and because *the hardness of the whole arises from the hardness of the parts*, [my italics] we therefore justly infer the hardness of the undivided particles, not only of the bodies we feel but of all others'. He goes on to apply this principle to the whole set of qualities which he needs to characterize bodies by the primary properties. 'The extension, hardness, impenetrability, mobility and inertia of the whole,' he says, 'result from the extension, hardness, impenetrability, mobility and inertia of the parts; and hence we conclude the least particles of all bodies to be all extended, and hard, and impenetrable, and movable, and endowed with their proper inertia.'

It is instructive in considering this general conceptual system to bring out the reasons why Newton felt himself obliged to look for a cause of gravity. On the face of it, since gravity is a universal phenomenon, a power inherent in all bodies whatsoever, it would look as if its universality would be sufficient to make a claim for its primacy, and hence by tacitly including it among the basic items listed in the general conceptual system, remove the necessity for an explanation of it. This course he explicitly rejects, for he says in his second letter to Bentley, 'You sometimes speak of Gravity as essential and inherent to Matter. Pray do not ascribe that notion to me; for the Cause of Gravity is what I do not pretend to know, and therefore would take more time to consider it.'

The reason why Newton could not treat gravity as a basic property of matter seems to me to lie in the fact that while gravitational action is invariant in one sense, that is cannot be affected by chemical, magnetic or calorific changes and obeys a known law very exactly, it is not invariant in the Newtonian sense, since it admits of intensification and remission of degree, depending, of course, on the inverse square of the distance of the point at which it acts from the body which is the 'source' of the gravitational action. It cannot therefore be a primary quality. Newton makes this point in a passage in his third letter to Bentley in which he says, 'It is inconceivable, that inanimate brute Matter should, without the Mediation of something else, which is not material, operate upon, and affect other Matter without mutual Contact, as it must be, if Gravitation in the sense of Epicurus, be essential and inherent in it. . . . Gravity must be caused by an Agent acting constantly according to certain Laws; but whether this Agent be material or immaterial I have left to the consideration of my Readers.' The reason Newton gives for doubting that gravity is 'essential and inherent in matter' is the inconceivability of action at a distance, but the sort of answer which he would regard as satisfactory to the question of the nature of gravity is 'an Agent acting *constantly*' and the 'constantly' is the key word here, for that would be a Newtonian invariance, and that Agent a primary property of matter.

THE GENERAL CONCEPTUAL SYSTEM AS A DETERMINANT OF METHOD

Two simple pictures of scientific method tend to develop from an insistence on an empirical basis for knowledge. One is the positivist picture which suggests that speculative activity is irrelevant to scientific success, the other, the inductivist picture, that all scientific knowledge is reached by the generalization of discoveries of particular fact. The culmination of the mechanical world picture in Newton's science we have seen as the final adoption of a particular general conceptual system, the acceptance of which determines both the direction in which the analysis of phenomena should proceed and the content which must be achieved to make an explanation acceptable. It seems also to be true, and this can be illustrated with Newton, that the adoption of a certain general conceptual system determines method too.

Two methods of forming hypotheses are used by Newton. They are both included under the general plan of forming an idea in the contemplation of the particular and rendering it 'general by induction'. But the 'rendering general' is, if we pay attention to all the Rules in which Newton sums up his philosophy of science, accomplished in two distinct ways. Consider first the generalization of particular instances of terrestrial gravitational attraction to a universal cosmological gravity. One might say that this is to generalize by *enlargement*. To claim that gravitational attraction is universal in this way is to make an hypothesis but it is not to invent anything, for the hypothesis contains no more content than the particulars from which it derives. It differs from the particulars only in generality of application of that content. It is to treat apples and planets as all *alike* in a certain respect.

But if we consider Newton's Rule III we find there another way of making hypotheses. The Primary properties of gross bodies are also assigned to minute bodies; and in the corresponding perceptual situation, the primary qualities of bodies which are and can be observed are also assigned to others too small to be observed. One might say that this is to generalize by *division*.

Science, and Newton's science is no exception, is full of speculative hypotheses, whose existence and use can hardly be justified on either positivist or empiricist grounds. But if we admit both generalization by enlargement, which will not justify speculative inventions of entities, and generalization by division which will, not only do we make intelligible the speculative methods widely and successfully used in science, but we also remove the appearance of contradiction between the penultimate and last paragraphs of Newton's General Scholium. In the penultimate paragraph he says that he does not venture hypotheses, when faced with a particular difficulty to be explained, while in the last paragraph he indulges in a staggering proliferation of hypothetical entities, processes and mechanisms, quite clearly with the intention of suggesting directions for future research. If we regard his remark against the making of hypotheses as designed to rule out hypotheses reached neither by enlargement nor division, then the subtle spirits and aethers of the speculative side of Newton escape the prohibition provided that they are reached by the process we have called generalization by division.

This exegesis removes the appearance of methodological contradiction at the expense of instituting an apparent contradiction of substance in the Newtonian system as a whole. The aethereal hypothesis is to be understood on this exegesis as a generalization by division, and hence the insensible particles of the aether *must* necessarily have all the primary properties argued for in Rule III since by division no primary properties can disappear. But after arguing for extension, hardness, impenetrability, mobility, inertia and divisibility Newton finally argues for the universality of gravitation though not for the primacy of that phenomenon. But now it seems that an aether whose particles are endowed with all the universal properties must therefore be endowed with the property of universal gravitation, if they are endowed with any. Hence it follows that the aether hypothesis, if it is a legitimate hypothesis in accordance with Rule III, cannot be used to explain universal gravitation, since it is itself endowed in its ultimate particles with gravitation.

But is Newton committed to the view that a property associated with primary properties is universally found with those properties? One is inclined to think that it must be so, but it is not easy to find any reason why one feels this. The argument for the universality of gravity in division is incidental to the argument that it is independent of form. The whole argument runs as follows: 'Universally, all bodies about the earth gravitate towards the earth; and the weights of all at equal distances from the earth's centre, are as the quantities of matter which they severally contain. This is the quality of all bodies within the reach of our experiments; and therefore (by Rule III) to be affirmed of all bodies whatsoever. If the aether, or any other body, were either altogether void of gravity, or were to gravitate less in proportion to its quantity of matter, then, because (according to *Aristotle, Descartes*, and others) there is no difference between that and other bodies in mere form of matter, by a successive change from form to form, it might be changed at last into a body of the same conditions with those which gravitated most in proportion to their quantity of matter . . . and therefore the weights would depend upon the forms of bodies, and with those forms, might be changed: contrary to what was proved in the preceding corollary.'

But in the Rules some of the force of this argument is withdrawn. Newton, after citing the universal attraction of all heavenly bodies, goes on to say . . . 'we must in consequence of this rule [i.e. the joint rule of generalization by division and enlargement] universally allow that all bodies whatsoever are endowed with a principle of mutual gravitation. For the argument from appearances concludes with more force for the universal gravitation of all bodies than for their impenetrability . . . not that I affirm gravity to be essential to bodies.' Hence, since universal gravity is not, as we saw, a primary property, there is no inconsistency in proposing to explain the phenomena of gravity in terms of an aether which gravitates, provided that only the primary properties of that aether figure in the explanation.

A Summary of the Arguments

The various forms of derivation and justification of the real properties of matter can now be brought together. The problem is not that of justifying the atomic hypothesis that matter is discontinuous, but that of justifying a certain assignment of properties to the ultimate divisions of matter. Gassendi, having argued for the logical independence of the concepts of matter and space, had made the atomic hypothesis respectable. But for the full justification of a corpuscularian philosophy a way or ways of determining what are the essential properties of matter is required.

Philosophers have been inclined to treat the argument for the distinction between primary and secondary qualities, and hence primary and secondary properties, as ways of justifying a certain analysis of experience, which, it is implied, might have been made by anyone at any time who examined the nature of experience and of our knowledge with sufficient attention. We should now be ready to abandon this idea, even for Locke, the philosopher whose way of putting his argument lends most colour to this view. My own account of my long example of the way a general conceptual system is justified, is, it should now be clear, designed to bring out a quite different aspect of this part of the philosophy of empiricism. It seems to me that if we consider the arguments of the scientists and the philosophers together we are drawn to the view that what is at issue is the assignment of that set of properties to the cor- puscles which will suffice for a science of matter in inter- action with itself. Incidentally, in eliminating, say, colour from this set of properties we assign the colour of a body to human sensibility and not to the body; or rather we say that what colour is in the body is not what it appears to be in human consciousness. This is why it is essential to Locke to distinguish ideas of qualities (what I have called just qualities) from the qualities themselves (what I have called properties). Berkeley was quite right to assert that all qualities are, as far as perception goes, on a par, but it should now be clear that with respect to the cogency of Locke's attempt at the task of justifying a general conceptual system it is quite irrelevant. To put this in another way; what constructors of the Newton- ian general conceptual system were in search of was a model in the perceptual world for the corpuscles of the scientific picture of reality.

The success of the mechanical sciences was at least *prima facie* evidence that the bulk, figure, motion and arrangement of the particles should be the basis of the scientific picture of physical reality. But more than pragmatic success of a system is needed to overthrow a rival general conceptual system. It was clearly felt at the time that reasons were needed, as Locke's account of himself as an under labourer testifies.

As I see it we are presented with three different ways of justifying the Newtonian g.c.s.; three different sorts of reasons which would make the corpuscularian philosophy acceptable. For the purposes of what follows I shall call them the *general*, the *empirical* and the *analytical* method.

(1) The general method: exemplified by the reasoning of Boyle's *Origin of Forms and Qualities*. In that work we are presented with a number of reasons, derived largely from experiment and observation, for accepting the general principle.

Quantitative change produces qualitative change.

From this proposition the Newtonian general conceptual system can be reached by the following moves:

(*a*) Quantitative change, however initiated, is change in the bulk, figure, motion and arrangement of corpuscles.

(*b*) Changes of this kind can lead only to other bulks, figure motions and arrangements of corpuscles.

(*c*) Qualitative change must then be, in the bodies where we observe it, actually quantitative change, that is change in quality must be the result, on human sensibility, of change in the bulk, figure, motion and arrangement of corpuscles.

(*d*) If it can be shown that every known quality can be changed by quantitative change, then the sciences need only use bulk, figure, motion and arrangement of the corpuscles as properties of matter.

(2) The empirical method: exemplified by Bacon's programme as set out in the *Novum Organon* and the *Valerius Terminus*. The perceptual world allows no clear distinction between Natures as to their Reality. What are the most real Natures will emerge from the empirical investigations of science. Those Natures which occur as the end products of tabular inductions are the Forms of Natures, and the total set of most real Natures are, for the purposes of science and technology, the fundamental properties of matter. To use a more modern example than Bacon; if we have found reason to believe that light is an electro-magnetic radiation, that Radiant Heat is an electro-magnetic radiation, that Radio is an electro-magnetic radiation, then Light, Radiant Heat and

Radio are three Natures having the same Form. Consequently electro-magnetic radiation is the most real Nature and consequently is a real feature of the world of which Light and Heat are the effects in organisms and Radio the effect on a particular kind of instrument. It seems to me quite consistent with Bacon's general views that the most real Nature should be a theoretical entity, property or process, only pointed to by the survivors of tabular inductions.

(3) The analytical method: exemplified, with differences, by the reasoning of Galileo in *Il Saggiatore* and Locke in the *Essay*. For both the task is to produce arguments and reasons for distinguishing one by one those perceptual qualities which are the product of mechanical action and human sensibility from those which are mechanical actions alone. The difference between their arguments, as I see it, reduces to this: that Galileo took it for granted that mechanical action was immediately recognizable and the task simply that of marking off those perceptual qualities dependent upon human sensibility, while Locke felt obliged to try to show by the argument of the perceptual counterpart that mechanical action can be recognized in sensation and perception, and hence that mechanical properties can be distinguished from those qualities which are the product of mechanical properties and human sensibility.

Acceptance of the doctrine of the corpuscularian philosophy, that matter is that which is defined by the primary properties, not only determines the details of a general conceptual system and hence the acceptable form for explanations, but also the details of acceptable scientific method. Denial of the doctrine of matter should have consequential effects, one would expect, upon what one believes to be the scope and capacity of scientific investigation. This study will be rounded off by a very brief examination of such a denial and the consequences drawn from it for the understanding of the nature of science. Berkeley's *Principles of Human Knowledge* contain the most radical and powerful attack upon Newtonian science that appeared before Mach's critical work in the nineteenth century. It is with Berkeley's attack upon the concept of matter that I shall complete my exposition of the Newtonian system.

Berkeley's criticisms are fundamental and comprehensive. His argument moves from an attack on the Lockean analysis of experience through a consequent denial of the existence of Newtonian matter to a strictly positivist account of the form and function of natural laws. He saw that science as it had been developed by Newton depended ultimately on the propriety of the distinction between primary and secondary properties, since this freed matter from identification with its effects. It gave it the capacity for independent existence. But there is no matter, Berkeley argues. There are in our experience only ideas ('idea' in Berkeley's usage means roughly what we should call 'sensation' and what Hume called 'impression'). 'And' he says,* 'as several of these [ideas] are observed to accompany each other, they come to be marked by one name, and so to be reputed as one thing . . . collections of ideas constitute a stone, a tree . . . and the like sensible things.' From here it is a short step to Berkeley's most characteristic doctrine. 'The Table I write on exists, that is I see it and feel it.'† We cannot deduce that there are independent primary and secondary properties since ideas, according to Berkeley are identical with properties and qualities both. For this conclusion he adduces several arguments two of which are important. Primary ideas, he argues, are inseparable from secondary ideas, for example it is not possible to conceive an extension without a colour. Hence, if we accept Locke's view that secondary ideas are in the mind and not copies or resemblances of the corresponding properties in objects, the same must be true of primary ideas or ideas of primary properties. So just the same status must be accorded to all our ideas. In case we think some form of Galileo's contrast argument depending upon the invariance of primary properties will rescue the distinction,

* In the *Principles*, Pt. 1, I.   † *Principles*, Pt. 1, IX.

Berkeley makes use of the argument I have already used in discussing Galileo's quasi-definition of matter, that there is as much relativity and variability among ideas of primary as among ideas of secondary properties.

Furthermore, Berkeley argues,* if only ideas do exist then even if we admitted groups of ideas of primary properties to some special status this would not help revive the doctrine of matter in the form needed by Newtonian science, since 'the very being of an idea implies passiveness and inertness', therefore 'it is impossible for an idea to be the cause of anything', in particular ideas of primary properties which indeed for Berkeley *are* the primary properties, cannot make up a causal background to experience. He concludes that the doctrine that our sensations 'are the effects of powers resulting from the configuration, number, motion and size of corpuscles must certainly be false'.

If this view is adopted and independent matter dropped from our ontology a different view of the nature of science forces itself on us. We still find that 'the ideas of sense are more distinct, . . . have steadiness, order and coherence . . . , are in a regular train or sequence,'† the phenomena which forced us to postulate matter as the fundamental invariant. But, according to Berkeley, the cause of the regularity which we should now see to be entirely among ideas must be a mind or spirit. The 'set of rules or established methods, wherein the mind we depend on excites in us the ideas of sense, are called the laws of nature, and these we learn by experience, which teaches us *that such and such ideas are attended with such and such other ideas, in the ordinary course of things*.' (my italics).

On this view the purpose of the laws of nature cannot be, among other things, to assist understanding but can only be for the prediction of the sensations we are likely to experience following upon the occurrence of some other sensations. In XLIV he says, 'In strict truth the ideas of sight, when we apprehend by them distance and things placed at a distance,

---

* *Principles*, Pt. 1, XXV.   † *Ibid*. Pt. 1, XXX.

do not suggest or mark out to us things actually existing at a distance, but only admonish us what ideas of touch will be imprinted in our minds at such and such distances of time, and in consequence of such and such actions.' Indeed, the proper function of natural philosophy is not to find causes but signs. Natural philosophy does not, according to Berkeley, use the concept of corporeal substance or matter, but actually uses 'figure, motion and other qualities which are in truth no more than mere ideas, and therefore cannot be the cause of anything' (L). When we do find some regularity in our experience Berkeley claims that the 'connexion of ideas does not imply the relation of cause and effect, but only of a mark or sign of the thing signified. The fire which I see is not the cause of the pain I feel upon my approaching it, but the mark that forewarns us of it . . . it is the searching after, and endeavouring to understand, those signs (this language if I may so call it) . . . that ought to be the employment of the natural philosophers.' So without the concept of matter what the laws of nature come down to in the last analysis are rules for working out what sensations are the signs of what other sensations. As for the penetration of nature in depth, to realms beyond our capacity to perceive, this is, according to Berkeley, a nonsensical proposal, since there is nothing but what we perceive. So we find in Berkeley's negative attitude to matter a ground and source of a positivism of a very fundamentalist kind.

What we have seen, however, of the corpuscularian philosophy does suggest that Berkeley's arguments are not really fatal to it. They are fatal to the suggestion that the corpuscularian philosophy is an analysis of experience, an epistemology. This we have seen it cannot be supposed to be. The corpuscularian philosophy is a general conceptual system—and as such it is an invention. But like that of Aristotle, and unlike that of modern physics, it drew its fundamental concepts from perception. What Berkeley did not grasp was that matter is a theoretical concept, though no doubt its main features are drawn from the perceivable properties of such stuffs as cheese and water. Its justification though does not consist in the impossible task of demonstrating that perceptual experience forces the primary properties of matter upon us, but in its power to unify and underwrite the laws of the superficial relations of phenomena.

INDEX

THE END